STEPPING

INTO THE WORLD OF

BEEKEEPING

A Thorough Initiation into
Sustainable Bee Farming

BEA HARRISON

TABLE OF CONTENTS

INTRODUCTION

Welcome to the world of beekeeping! Whether you're looking to start your own hive, or simply learn more about these fascinating creatures, you've come to the right place.

What is Beekeeping?

Beekeeping, also known as apiculture, is the practice of managing honey bee colonies, typically in hives, by humans. It is a form of agriculture that has been practiced for centuries, and is used to produce honey, beeswax, pollen, propolis, and other bee products.

Beekeeping is a complex and rewarding activity that requires knowledge, skill, and dedication. Beekeepers must understand the biology and behaviour of honey bees, and be able to recognize and manage diseases and pests. They must also be able to identify and manage the various bee products, and be able to construct and maintain hives and other beekeeping equipment.

Beekeeping is a practice that has been around for thousands of years, and for good reason. Bees are some of the most important creatures on our planet, playing a vital role in pollinating the plants that provide us with food, oxygen, and beauty.

But beekeeping isn't just about the benefits it provides to the environment and our lives. It's also a rewarding and fulfilling hobby that can bring you closer to nature and give you a new appreciation for the natural world. There's something truly special about tending to a hive of bees, watching them work together and create something amazing.

Of course, beekeeping is not without its challenges. As with any animal or plant, bees require a certain level of care and attention to thrive. But with the right knowledge and equipment, beekeeping can be a relatively easy and enjoyable hobby, even for beginners.

But before we dive into the details, let's take a moment to explore the history of beekeeping and the tips and strategies for getting the most out of this guide.

Brief History of Beekeeping

Beekeeping is a practice that has been around for thousands of years. From ancient Egypt to modern-day America, people have recognized the importance of bees in pollinating plants and producing honey. The history of beekeeping is a rich and fascinating one, spanning millennia and encompassing a variety of cultures and traditions.

The earliest evidence of beekeeping comes from rock paintings in Valencia, Spain, that date back to around 7000 BCE. These paintings depict humans collecting honey from wild bee colonies, and may represent some of the earliest known examples of human interaction with bees.

Over time, humans began to develop more advanced techniques for collecting honey and managing bee colonies. One of the earliest methods involved burning or smoking bees out of their hives in order to access the honey. Later, humans developed more sophisticated techniques, such as using hollow logs or baskets to attract bees and collect honey.

The ancient Egyptians were among the first civilizations to develop beekeeping as a formal practice. Honey was a valuable commodity in ancient Egypt, and was used for a variety of purposes. In fact, the ancient Egyptians even considered honey to be a sacred food, and used it in religious rituals and offerings. The Egyptians developed several advanced techniques for beekeeping, including the use of hives made of reeds or clay.

The ancient Greeks and Romans also practiced beekeeping, and were among the first to recognize the importance of bees in pollinating crops. The Greek philosopher Aristotle wrote extensively about bees and their behavior, and the Roman naturalist Pliny the Elder wrote about the medicinal properties of honey and beeswax.

During the Middle Ages, beekeeping became a popular practice in Europe, with monks and other religious orders taking up the practice as a way to produce honey and other bee products. By the 16th century, beekeeping had become a widespread practice throughout Europe, with various techniques and hives being developed to improve the efficiency and productivity of honeybees.

One of the most significant developments in the history of beekeeping came in the 19th century, with the invention of the movable frame hive. Prior to this invention, beekeepers had to destroy the hive in order to harvest the honey, which was a time-consuming and inefficient process. With the movable frame hive, beekeepers were able to remove individual frames of honey without disturbing the rest of the colony, making it easier to manage and harvest honey from the hive.

Today, beekeeping is a popular hobby all over the world, with millions of people keeping hives in their backyards, on rooftops, and in rural areas. While many of the techniques and tools used in beekeeping have evolved over the centuries, the basic principles remain the same: creating a safe and healthy environment for bees to thrive, and harvesting the honey and other products they produce.

In addition to providing honey and beeswax, beekeeping plays an important role in pollinating plants and supporting the environment. Bees are essential pollinators, helping to ensure that plants reproduce and grow. Many plants wouldn't be able to reproduce without bees, which would harm ecosystems. By keeping bees, beekeepers are helping to support the environment and ensure the health of the ecosystem.

We'll walk you through all you need to know in this detailed guide to get started with beekeeping, from the basics of bee anatomy and behavior to the ins and outs of hive maintenance and honey harvesting. We'll also explore the fascinating history of beekeeping, and the many cultural and technological developments that have shaped this ancient practice over the centuries.

Whether you're a seasoned beekeeper looking to expand your knowledge, or a beginner just starting out, this guide is for you. With the right knowledge and equipment, beekeeping can be a relatively easy and enjoyable hobby, even for beginners.

We hope that this book will provide you with the knowledge and skills necessary to become a successful beekeeper.

How to use this guide?

Before we get started, it's important to understand how to use this guide to get the most out of it. Here are some tips to help you make the most of the information provided in this book:

1. **Start at the beginning** - While you may be eager to dive right into hive maintenance or honey harvesting, it's important to begin reading the book at the beginning and to proceed through the chapters in order. This will provide you with a solid foundation of knowledge and make sure you comprehend the fundamentals before going on to more advanced topics.

2. **Take notes** - Beekeeping is a complex and nuanced practice, and there is a lot of information to absorb. To help you remember important details, consider taking notes as you

read through the book. You can use a notebook, a digital document, or any other method that works best for you.

3. **Ask questions** - If you're unsure about a particular concept or technique, don't be afraid to ask questions. Beekeeping can be a challenging hobby, and it's normal to feel overwhelmed or confused at times. You can reach out to beekeeping groups, forums, or mentors to get answers to your questions.

4. **Practice** - Beekeeping is a hands-on activity, and the best way to learn is by doing it. As you read through the chapters, try to apply the concepts and techniques to your hive. Try out several techniques and tools to see one suits you the best.

5. **Be patient** - Beekeeping is not a hobby that can be mastered overnight. It takes time and practice to develop the skills and knowledge needed to be a successful beekeeper. Be patient with yourself and your bees, and enjoy the process of learning and growing as a beekeeper.

So, if you're ready to dive into the world of beekeeping, grab your protective gear and let's get started! By following the tips mentioned above and working your way through the chapters of this guide, after a short period of time, you'll be well on your way to becoming a successful beekeeper.

We wish you the best of luck in your beekeeping journey!

CHAPTER
I
Benefits and
Importance of Beekeeping

Benefits of Beekeeping

Beekeeping is a long-established tradition that dates back thousands of years. From the earliest days of human civilization, people have recognized the importance of bees in pollinating plants and producing honey. Today, beekeeping is more popular than ever, with millions of people around the world keeping bees in their backyards, on rooftops, and in rural areas.

But why is beekeeping such a popular hobby, and what are the benefits of keeping bees? In this introduction, we'll explore some of the many benefits of beekeeping, from the impact it has on the environment to the health benefits of honey and other bee products.

One of the most important benefits of beekeeping is that it helps to promote biodiversity. Bees are essential pollinators, and they are responsible for pollinating a large percentage of the world's food crops, including fruits, vegetables, and nuts. Without bees, many of the foods that we eat on a daily basis would not exist. In fact, it's estimated that one third of the world's food supply depends on honeybee pollination. By keeping bees, you are helping to ensure that these important pollinators are able to do their job. This helps to ensure that food crops are able to grow and thrive, which in turn helps to ensure that people have access to a variety of healthy foods.

In addition to their role in pollination, another benefit of beekeeping is that it can help to improve your health. Honey is a natural sweetener that can be used in a variety of ways, from baking and cooking to adding flavor to tea and other beverages. It is a natural source of antioxidants, which can help to reduce inflammation and improve overall health and also rich in other nutrients, making it a healthy alternative to processed sugar. Beeswax, on the other hand, can be used to make candles, soaps, and other beauty products. Both honey and beeswax have been used for centuries for their medicinal properties, and are still valued for their healing benefits today.

Additionally, bee pollen is a great source of vitamins and minerals, and it can help to boost your immune system.

Beekeeping can also be a source of income, especially for those who are willing to sell their honey and other bee products. While it may take some time to establish a successful beekeeping operation, those who are dedicated and willing to put in the work can turn their hobby into a profitable business.

But perhaps one of the greatest benefits of beekeeping is the sense of connection it provides to nature. Tending to a hive of bees can be a soothing and peaceful experience that aids in slowing you down and appreciate the natural world around you, and it can be a great way to relax and unwind. Watching the bees as they work together to build their hive and collect nectar is a fascinating experience, and can provide a sense of fulfillment that is hard to find in our fast-paced, modern world.

In addition to the benefits to the environment and our physical health, beekeeping has also been shown to have mental health benefits. Research have demonstrated that being in nature and participating in activities like beekeeping, lessen your tension and stress, and improve mood and overall well-being.

The role of bees in pollination and the environment

Bees are among the most important pollinators in the world, playing a vital role in sustaining many plant species and ecosystems. In fact, it is estimated that bees are responsible for pollinating more than 80%

of the world's flowering plants, including crops that provide one-third of the world's food supply.

Pollination is the process of moving pollen from a flower's male to female parts, which enables the plant to generate seeds and fruit. Bees are particularly effective pollinators because they have specialized body parts and behaviors that help them collect and transfer pollen from one flower to another.

The process of pollination also has important ecological benefits. For example, it helps to maintain plant biodiversity by facilitating the reproduction of a wide variety of plant species. This, in turn, supports the survival of many animal species that rely on plants for food and habitat.

Bees also contribute to the economic well-being of many communities by playing a critical role in pollinating crops that are important for human consumption, such as fruits, vegetables, and nuts. In fact, it is estimated that the global value of crops pollinated by bees is in the hundreds of billions of dollars each year.

Unfortunately, many bee populations are currently in decline due to a variety of factors, including habitat loss, pesticide exposure, and disease. This decline has serious implications for both the environment and the economy, as it could lead to reduced biodiversity, lower crop yields, and higher food prices.

To address this issue, there are many efforts underway to promote bee conservation and support sustainable beekeeping practices. These efforts include creating new habitats for bees, reducing pesticide use, and promoting the use of sustainable agricultural practices that support healthy bee populations.

Overall, the role of bees in pollination and the environment is crucial, and it is important that we take steps to protect and support these important pollinators.

The importance of beekeeping for agriculture

Beekeeping is an ancient practice that has been used for thousands of years to produce honey and other hive products, such as beeswax and royal jelly. However, beekeeping also has important implications for agriculture, as it can play a crucial role in pollinating crops and promoting sustainable agricultural practices.

In this section, we will explore the importance of beekeeping for agriculture and examine the various ways in which beekeeping can support sustainable farming practices and contribute to global food security.

Pollination and Crop Yields

One of the most important roles of bees in agriculture is their ability to pollinate crops. Pollination is essential for crop production, as it allows plants to produce fruits and seeds. Without pollination, many crops would not be able to produce the yields necessary to meet global food demand. For example, studies have shown that bee pollination can increase crop yields by up to 70% for some crops, such as almonds.

In addition, bee pollination can also improve the quality of crops. For example, research has shown that bee pollination can increase the size, shape, and color of fruits, such as strawberries and blueberries.

Sustainable Agriculture

Beekeeping can also support sustainable agricultural practices by promoting biodiversity and reducing the need for chemical fertilizers and pesticides. When bees pollinate crops, they also help to support the health and diversity of local ecosystems by facilitating the reproduction of a wide variety of plant species.

In addition, beekeeping can help to reduce the use of chemical fertilizers and pesticides by promoting natural pest control methods. For example, some beekeepers use hives to attract and control pests, such as mites, that can damage crops. This can help to reduce the need for chemical pesticides and promote sustainable farming practices.

Economic Benefits

Beekeeping can also provide important economic benefits for farmers and rural communities. In addition to producing honey and other hive products, beekeeping can also generate income through the sale of pollination services.

For example, many farmers hire beekeepers to bring hives to their farms during the blooming season to pollinate their crops. This can increase crop yields and improve crop quality, which can result in higher profits for farmers. In addition, beekeeping can also provide income for rural communities through the sale of honey and other hive products.

Challenges Facing Beekeeping

Despite the many benefits of beekeeping for agriculture, there are also a number of challenges facing beekeepers and the beekeeping industry. One of the biggest challenges is the decline in bee populations in many parts of the world. This decline is due to a number of factors, including habitat loss, pesticide exposure, and disease.

Habitat loss is a major threat to bee populations, as it can limit the availability of food and nesting sites. As natural habitats are destroyed to make way for agriculture and urban development, bee populations can decline, which can have serious implications for pollination and food security.

Pesticide exposure is another major threat to bee populations. Many pesticides can be toxic to bees, and exposure to these chemicals can lead to reduced reproductive success and increased mortality. In addition, some pesticides can also have long-term effects on the health and behavior of bees, which can lead to declines in population size and reproductive success.

Disease is also a significant threat to bee populations, as many diseases can be transmitted from hive to hive and can spread rapidly throughout a population. Common bee diseases include American foulbrood, European foulbrood, and chalkbrood, all of which can have serious consequences for hive health and productivity.

Conservation and Sustainable Beekeeping Practices

To address the challenges facing beekeeping, there are many efforts underway to promote bee conservation and support sustainable beekeeping practices. These efforts include creating new habitats for bees, reducing pesticide use, and promoting the use of sustainable agricultural practices that support healthy bee populations.

Developing new bee habitats is a crucial part of bee conservation efforts. This can include planting bee-friendly plants and flowers that provide a source of nectar and pollen for bees. It can also involve creating nesting sites for bees, such as artificial hives or structures that provide shelter and protection.

Reducing pesticide use is another important strategy for supporting healthy bee populations. This can involve promoting the use of organic and sustainable farming practices that limit the use of pesticides and other harmful chemicals. It can also involve educating farmers and beekeepers about the risks associated with pesticide use and promoting alternatives to chemical pest control.

Promoting sustainable beekeeping practices is also an important component of bee conservation efforts. This can involve educating beekeepers about best practices for hive management and disease prevention. It can also involve promoting the use of bee-friendly materials and techniques, such as natural honeycomb and low-impact harvesting methods.

Policy and Regulation

In addition to conservation and sustainable beekeeping practices, policy and regulation can also play an important role in supporting healthy bee populations and promoting sustainable agriculture. For example, many countries have developed regulations that limit the use of pesticides that are harmful to bees, such as neonicotinoids. Some countries have also developed programs to support beekeepers and promote sustainable beekeeping practices.

At the international level, there are also a number of initiatives underway to support bee conservation and sustainable agriculture. For example, the Food and Agriculture Organization (FAO) of the United Nations has launched a Global Action on Pollination Services for Sustainable Agriculture, which aims to promote sustainable agriculture and support healthy bee populations through research, policy development, and capacity building.

Therefore, Beekeeping is a critical component of sustainable agriculture, providing important pollination services and supporting the health and diversity of local ecosystems. However, beekeeping also faces a number of challenges, including declines in bee populations due to habitat loss, pesticide exposure, and disease.

To support healthy bee populations and promote sustainable agriculture, it is important to promote conservation and sustainable beekeeping practices, as well as develop policies and regulations that support healthy bee populations and promote sustainable agriculture.

By working together to support bee conservation and sustainable agriculture, we can help to ensure the health and well-being of both bees and humans for generations to come.

Choosing the right type of beekeeping for you

Beekeeping comes in a variety of forms, each with special benefits and traits of its own. In this section, we will explore the different types of beekeeping and provide guidance on how to choose the right one for your needs and preferences. Some of the most common types of beekeeping include:

Types of Beekeeping

1. **Backyard Beekeeping** - Backyard beekeeping is the most common type of beekeeping and involves keeping a small number of hives in a backyard or other small area. This is an excellent option for hobbyists and those interested in producing honey and other hive products for personal use.

2. **Urban Beekeeping** - Urban beekeeping involves keeping hives in an urban or suburban environment, often on rooftops or in other small spaces. This is an excellent option for those who live in a city or other urban environment and are interested in promoting sustainability and supporting local ecosystems.

3. **Commercial Beekeeping** - Commercial beekeeping involves keeping large numbers of hives and producing honey and other hive products for sale. This is an excellent option for those interested in starting a small business or working in the agriculture industry.

4. **Natural Beekeeping** - Natural beekeeping involves keeping bees in a way that mimics their natural habitat and behavior, using minimal intervention and allowing the bees to build their own combs and colonies. This is an excellent option for those interested in promoting sustainable beekeeping practices and supporting healthy bee populations.

When choosing the right type of beekeeping for you, it is important to consider several factors, including your experience level, your goals for beekeeping, and your available resources.

Experience Level - If you are new to beekeeping, it is important to start with a small number of hives and gradually build your skills and experience over time. Backyard beekeeping is an excellent option for beginners, as it allows you to start small and gradually increase the size of your operation as you become more comfortable and confident in your skills.

Goals for Beekeeping - Your goals for beekeeping will also play a role in determining the right type of beekeeping for you. If you are interested in producing honey and other hive products for personal use, backyard beekeeping or urban beekeeping may be the best option for you. If you are interested in starting a small business, commercial beekeeping may be a better fit. If you are interested in promoting sustainable beekeeping practices, natural beekeeping may be the right choice.

Available Resources - The resources you have available, including time, space, and money, will also play a role in determining the right type of beekeeping for you. Backyard beekeeping and urban beekeeping require relatively little space and investment, making them an excellent option for those with limited resources. Commercial beekeeping requires a significant investment in equipment and infrastructure, and may be best suited for those with more resources and experience.

Choosing the right type of beekeeping for you is an important decision that should be based on your experience level, goals, and available resources. Backyard beekeeping, urban beekeeping, commercial beekeeping, and natural beekeeping are all excellent options, each with its own unique characteristics and benefits.

By giving considerable thought to your requirements and preferences, and by seeking out the advice and guidance of experienced beekeepers, you can choose the right type of beekeeping for you and enjoy the many benefits that beekeeping has to offer.

CHAPTER
II
Bee Biology and Behavior

Bee anatomy and physiology

The anatomy and physiology of bees are highly specialized and are essential for their role as pollinators. Understanding the external and internal anatomy of bees, as well as their nervous system, sensory systems, and behavioral and physiological adaptations, is essential for beekeepers, scientists, and anyone interested in the biology of bees.

External Anatomy

The external anatomy of bees is highly specialized and adapted to their role as pollinators. Bees are insects with a head, thorax, and abdomen as their three primary body sections. The mouthparts, antennae, and eyes are located in the head. The wings and legs are located in the thorax, while the digestive system and reproductive systems are located in the abdomen.

External Anatomy
of a Honey Bee

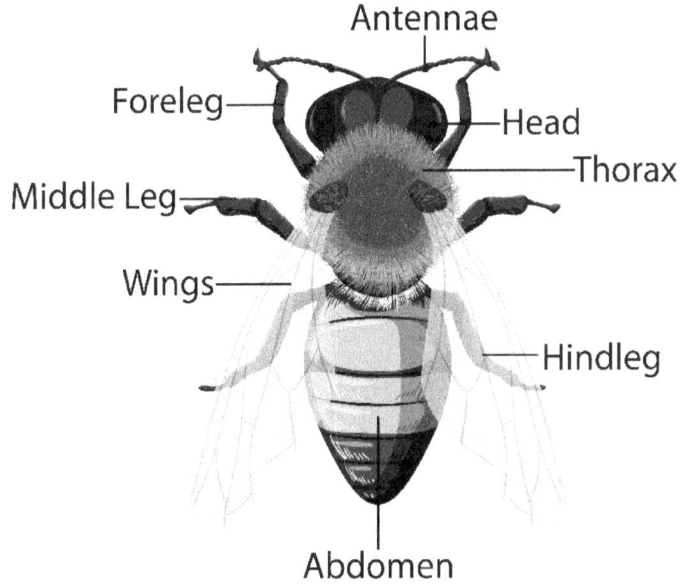

The eyes of bees are highly specialized and are composed of thousands of individual lenses that allow them to see in great detail and at high speeds. Bees also have three simple eyes, called ocelli, which are found on the top of the head and are used for navigation and flight stability.

The antennae of bees are also highly specialized and are used for a variety of functions, including touch, taste, and smell. The mouthparts of bees are designed for feeding on nectar and pollen and include a long, flexible tongue called a proboscis.

The wings of bees are another important feature of their anatomy. Bees have two pairs of wings that are used for flight and stability during pollination. The legs of bees are also highly adapted for their role as pollinators and contain structures that allow them to collect and transport pollen.

Internal Anatomy

The internal anatomy of bees is also highly specialized and is adapted to their unique physiology and behaviors. The digestive system of bees is composed of several parts, including the crop, the honey stomach, and the intestines. The crop is used to store nectar and other liquids, and the honey stomach is used to store and transport honey.

The reproductive system of bees is also highly specialized and includes the queen, drones, and worker bees. The queen is responsible for laying eggs, and drones are responsible for fertilizing the queen. Worker bees are in charge of collecting nectar and pollen, caring for the young, and maintaining the hive.

Nervous System

The nervous system of bees is also highly adapted to their unique behaviors and sensory systems. Bees have a centralized nervous system that includes a brain and ganglia that are distributed throughout the body. The brain of bees is highly specialized and contains regions that are responsible for processing sensory information, regulating behavior, and controlling movement.

Sensory Systems

The sensory systems of bees are also highly specialized and are essential for their role as pollinators. Bees have five main senses: sight, smell, touch, taste, and hearing. The eyes of bees are particularly important for their role as pollinators, as they are able to detect colors and patterns that are invisible to humans.

The sense of smell is also important for bees, as it allows them to locate and identify flowers and other sources of nectar and pollen. The sense of touch is used for a variety of functions, including communication, navigation, and flight stability. The sense of taste is used to detect and evaluate the quality of nectar and pollen.

Finally, the sense of hearing is used for communication and navigation. Bees are able to communicate with each other using a variety of sounds, including buzzing and waggle dances, which allow them to share information about the location of food sources and other important resources.

Behavioral Adaptations

In addition to their unique anatomy and physiology, bees have also evolved a variety of behavioral adaptations that allow them to function as highly efficient pollinators. For example, bees are able to communicate with each other using a variety of signals, including pheromones and sound.

Bees also have a highly structured social organization, with specialized roles for different members of the hive. The queen is responsible for laying eggs, drones are responsible for fertilizing the queen, and worker bees are responsible for collecting nectar and pollen, caring for the young, and maintaining the hive.

Another important behavioral adaptation of bees is their ability to navigate and forage over long distances. The position of the sun, the earth's magnetic field, and prominent features in the environment are just a few of the cues that bees use to navigate. They are also able to communicate the location of food sources to other members of the hive using complex dance behavior.

Physiological Adaptations

Bees have also evolved a variety of physiological adaptations that allow them to function as highly efficient pollinators. For example, bees are able to regulate their body temperature using a process called endothermy, which allows them to remain active even in cold or hot environments.

Bees are also able to produce wax for building the hive and honey for food storage. The wax is produced by glands on the underside of the abdomen, while the honey is produced by the workers from the nectar they collect from flowers.

Another important physiological adaptation of bees is their ability to detoxify and eliminate harmful substances from their bodies. Bees are exposed to a variety of toxins and pollutants in their environment, including pesticides and heavy metals, and have developed mechanisms for detoxification and elimination.

Therefore, by studying the anatomy and physiology of bees, we can gain a deeper understanding of their unique behaviors and adaptations, and develop strategies for promoting the health and well-being of these important pollinators.

Bee lifecycle and social organization

The life cycle and social organization of bees are complex and fascinating, reflecting their ability to thrive in a variety of environments and their crucial role in pollination.

Bee Life Cycle

The life cycle of bees begins with the hatching of an egg laid by the queen. The egg hatches into a larva, which is fed by the worker bees with a mixture of nectar and pollen. The larva undergoes several molts as it grows and develops, and eventually spins a cocoon around itself.

Within the cocoon, the larva undergoes metamorphosis, transforming into an adult bee. The adult bee chews its way out of the cocoon and emerges as a fully formed worker bee, drone, or queen bee.

The length of a bee's life depends on their function in the hive. Worker bees typically live for several weeks, while drones typically live for a few months. Queen bees can live for several years and are responsible for laying the eggs that give rise to the next generation of bees.

Egg Stage

The egg stage of the bee life cycle is the beginning of the process. The queen bee lays the eggs in the cells of the comb, which are hexagonal cells constructed by worker bees out of wax. The queen bee will lay either a fertilized or unfertilized egg, depending on whether she has mated with a drone. Fertilized eggs develop into female bees, while unfertilized eggs develop into male bees.

Larval Stage

After the egg hatches, the bee enters the larval stage. The larva is a small, white grub-like creature that is fed by worker bees with a mixture of nectar and pollen. During this stage, the larva undergoes several molts as it grows and develops.

The larva is housed in a wax cell constructed by the worker bees, and the cell is capped with wax when the larva is ready to spin a cocoon around itself.

Pupal Stage

The pupal stage is a period of metamorphosis during which the larva transforms into an adult bee. The pupa is housed in a cocoon spun by the larva, and the cocoon is constructed out of wax secreted by the worker bees.

During this stage, the bee undergoes a dramatic transformation, with the larval body structure completely rearranged into the adult bee body. The adult bee emerges from the cocoon and chews its way out of the cell.

Adult Stage

The adult stage is the final stage in the bee life cycle. The adult bee is responsible for collecting nectar and pollen, caring for the young, and maintaining the hive. The adult bee will live for a varying amount of time, depending on its role in the hive. Worker bees typically live for several weeks, while drones typically live for a few months. Queen bees can live for several years.

Social Organization of Bees

The social organization of bees is characterized by the division of labor among the individuals in the colony, as well as a sophisticated system of communication and a unique hive architecture.

The organization of bee societies can be traced back millions of years, and has evolved to be highly efficient and productive. The social organization of bees is divided into various categories, including the queen bee, worker bees, drones, brood, communication, division of labor, and hive architecture. Each of these categories is essential for the functioning of the colony, and they work together in a complex and interdependent manner.

A Queen Bee

The queen bee is the most significant individual in the colony, and her primary role is to lay eggs. The only reproductive female in the colony is the queen bee, and she is responsible for maintaining the unity of the colony by producing chemical signals that regulate the behavior of the other bees. The queen bee is larger than the other bees in the colony, and can live for several years.

When a new colony is formed, the first task of the worker bees is to select a new queen bee. They do this by constructing queen cells, which are larger than normal cells and are used to rear new queen bees. The first queen bee to emerge from her cell will usually kill the other developing queen bees in their cells, ensuring that she is the only queen bee in the colony.

The queen bee plays a crucial role in maintaining the social organization of the colony. She produces pheromones that help to keep the worker bees in a cooperative and productive state. These pheromones also help to suppress the development of other queen bees within the colony, ensuring that there is only one queen at a time.

Worker Bees

Worker bees are sterile females that perform various tasks within the colony, such as nursing the young, cleaning the hive, foraging for food, and defending the colony. Workers are smaller than the queen bee and have a shorter lifespan, typically only a few weeks to a few months.

The tasks performed by worker bees vary according to their age. Younger worker bees are responsible for nursing the brood and cleaning the hive, while older worker bees are responsible for foraging for food and defending the colony.

The division of labor among the worker bees is highly efficient, and ensures that each bee is able to specialize in a particular task. This division of labor helps to ensure that the colony is productive and able to meet the needs of all its members.

Drones

Drones are male bees whose primary role is to mate with the queen bee. They do not have stingers and do not perform any other tasks within the colony. While drones are smaller than the queen bee, they are bigger than worker bees.

Drones are produced by the queen bee during the spring and summer months, when the colony is growing and there is a need for new drones. Drones are usually expelled from the colony during the fall, when their primary role of mating with the queen bee is no longer necessary.

Brood

Brood refers to the immature bees in the colony, which include eggs, larvae, and pupae. The queen bee is responsible for laying eggs, and the worker bees take care of the brood by feeding and cleaning them.

The development of the brood is closely tied to the production of the queen bee's pheromones. The queen bee produces pheromones that inhibit the development of worker bee ovaries, ensuring that the workers remain sterile and focused on their tasks.

Communication

Bees have a complex system of communication that allows them to work together effectively. They use pheromones, chemical signals, and dances to convey information about food sources, the location of the hive, and the status of the queen bee.

One of the most fascinating forms of communication used by bees is the "waggle dance." This dance is performed by worker bees to communicate the location of food sources to other members of the colony. The dance involves the bee waggling its abdomen back and forth while moving in a figure-eight pattern. The as well as the length of the dance indicate the quality of the food source as well as its direction and distance.

Bees also use pheromones to communicate with each other. These chemical signals are produced by the queen bee and other members of the colony, and are used to coordinate the activities of the bees. Pheromones can be used to signal the location of the hive, the presence of a queen bee, and other important information.

Division of Labor

The colony is organized in a way that allows each bee to specialize in a particular task. For example, some worker bees are responsible for nursing the young, while others are responsible for foraging for food. This division of labor helps to ensure that the colony is efficient and productive.

The division of labor among the bees is not fixed, and can change according to the needs of the colony. For example, if there is a shortage of foragers, some worker bees may switch from their normal tasks to foraging.

Hive Architecture

Bees construct their hives out of wax, which they produce themselves. The hive is divided into various compartments, including brood cells, honeycomb, and storage cells for pollen and nectar.

The honeycomb is a particularly important part of the hive. It is used to store honey and pollen, as well as to provide a place for the queen bee to lay her eggs. The honeycomb is also used for communication, with bees using it to signal the location of food sources and other important information.

The structure of the hive is highly efficient, and is designed to minimize the amount of energy needed to maintain the colony. For example, the cells of the honeycomb are hexagonal in shape, which allows for maximum storage capacity while using the least amount of wax.

In conclusion, the social organization of bees is a fascinating example of how cooperation and division of labor can lead to the success of a complex society. The queen bee, worker bees, drones, brood, communication, division of labor, and hive architecture all work together to create a thriving colony.

The organization of bee societies is highly efficient and productive, and has evolved over millions of years to meet the needs of the colony. The division of labor among the bees ensures that each bee is able to specialize in a particular task, while the complex system of communication allows the bees to work together effectively.

The unique hive architecture of bees is also an important factor in their success. The honeycomb provides a highly efficient way to store honey and pollen, as well as to provide a place for the queen bee to lay her eggs. The structure of the hive is designed to minimize the amount of energy needed to maintain the colony, ensuring that the bees are able to thrive in their environment.

Overall, the social organization of bees is a testament to the power of cooperation and division of labor in creating a successful society. The organization of bee societies has fascinated scientists and beekeepers alike for centuries, and continues to be an area of ongoing research and discovery.

Understanding bee communication and behavior

Bee Communication

Bee communication is a fascinating and complex topic that has captured the attention of scientists and beekeepers alike. Bees use a variety of methods to communicate with each other, including pheromones, sounds, vibrations, and dances. Understanding bee communication is essential for anyone interested in these remarkable creatures.

Pheromones

Pheromones are chemical signals that are used by bees to communicate with each other. Bees produce a wide variety of pheromones, which can be used to signal the location of food sources, the presence of predators, and the status of the queen bee.

One of the most important pheromones produced by bees is the queen mandibular pheromone. This pheromone is produced by the queen bee and is used to regulate the behavior of the other bees in the colony. The queen mandibular pheromone inhibits the development of worker bee ovaries, ensuring that the workers remain sterile and focused on their tasks.

Bees also produce pheromones to signal the presence of food sources. For example, when a worker bee finds a good source of nectar or pollen, it will return to the hive and release a pheromone that signals the other bees to follow it to the food source.

Sounds

Bees also use sounds to communicate with each other. For example, when a worker bee finds a good source of food, it will buzz its wings rapidly to signal to the other bees in the colony that it has found a food source.

Bees also use sounds to communicate with each other during swarming. Swarming is a natural behavior exhibited by bees when a colony becomes too large. During a swarm, a sizable number of bees will depart the hive and establish a new colony somewhere else. The bees use sounds to communicate the location of the new colony and to coordinate their activities during the swarm.

Vibrations

Bees use vibrations to communicate information about the location of food sources and the presence of predators. For example, if a bee detects a predator near the hive, it may signal to other bees by vibrating its body.

Bees also use vibrations to signal the location of food sources. For example, when a worker bee finds a good source of pollen, it will return to the hive and vibrate its body in a particular way to signal the other bees to follow it to the food source.

Dances

One of the most well-known forms of communication used by bees is the waggle dance. This dance is performed by worker bees to communicate the location of food sources to other members of the colony. The dance involves the bee waggling its abdomen back and forth while moving in a figure-eight pattern. The direction and duration of the dance indicate the direction and distance of the food source, as well as its quality.

The waggle dance is highly organized and efficient, and allows bees to locate food sources quickly and effectively. The dance is also flexible, and can be adjusted to take into account changes in the location or quality of the food source.

Each of these communication methods is essential for the functioning of the colony and allows the bees to work together effectively to meet the needs of the colony.

Behavior of Bees

Bees exhibit a wide range of behaviors, including foraging, nursing, hive defense, swarming, and reproduction. Each of these behaviors is regulated by a complex system of pheromones and other chemical signals, and is crucial for the survival of the colony.

Foraging

Foraging is the process by which bees collect nectar, pollen, and other resources from the environment. Foraging behavior is highly organized and efficient, with bees using a variety of methods to locate and collect food.

When a worker bee finds a good source of food, it will return to the hive and perform the waggle dance to communicate the location of the food source to other members of the colony. Other worker bees will then follow the dancing bee to the food source and begin to collect nectar and pollen.

Nursing

Nursing behavior is exhibited by worker bees who are responsible for caring for the brood. This includes feeding and cleaning the larvae, as well as keeping the hive's temperature and humidity constant.

Nursing behavior is highly organized, with different worker bees performing different tasks depending on their age. Younger worker bees are responsible for feeding and cleaning the larvae, while older worker bees are responsible for maintaining the temperature and humidity of the hive.

Hive Defense

Bees are highly territorial and will defend their hive against predators and other threats. Hive defense behavior includes stinging, biting, and swarming in large groups to attack predators.

Hive defense behavior is highly organized, with different worker bees performing different tasks depending on their role in the colony. For example, guard bees are responsible for patrolling the entrance to the hive and detecting potential threats, while other worker bees are responsible for attacking predators that come too close to the hive.

Swarming

Swarming is a natural behavior exhibited by bees when a colony becomes too large. During a swarm, a large group of bees will leave the hive and form a new colony elsewhere.

Swarming behavior is regulated by a complex system of pheromones and other chemical signals. The queen bee produces pheromones that signal to the other bees that it is time to swarm, while worker bees use the waggle dance and other communication methods to locate a suitable site for the new colony.

Reproduction

Reproduction behavior includes the mating of the queen bee with drones and the laying of eggs. Reproduction behavior is closely tied to the production of pheromones by the queen bee, which help to regulate the behavior of the other bees in the colony.

When a queen bee is ready to mate, she will leave the hive and fly to a location where drones are flying. The queen bee will mate with several drones, and then return to the hive to begin laying eggs.

As we continue to learn more about the behavior and communication of bees, we gain a greater appreciation for the complexity and sophistication of these remarkable creatures. The study of the behavior and communication of bee is an ongoing area of research and discovery, and promises to yield many new insights into the behavior and biology of these important insects.

The role of the queen bee

Queen bees are the heart of the honeybee colony, and their importance to the colony cannot be overstated. As the only reproductive female, she is responsible for laying eggs, producing pheromones, and regulating the behavior of the other bees. In this section, we will explore the role of the queen bee in more detail.

The Queen Bee's Physical Characteristics

Queen bees are physically distinct from the other bees in the colony. They are larger in size and have a longer abdomen than the workers or drones. They also have distinctive markings on their bodies, which help the workers to identify them.

One of the most important physical characteristics of the queen bee is her reproductive system. She has fully developed ovaries, which allow her to lay thousands of eggs during her lifetime. Her body is also equipped with specialized glands that produce pheromones, which play a critical role in the organization and functioning of the colony.

Egg Laying

Laying eggs is the queen bee's main responsibility. A healthy queen bee can lay up to 2,000 eggs per day, and is responsible for maintaining the population of the colony. The queen bee lays her eggs in cells of the honeycomb, which are prepared by worker bees.

The queen bee can regulate the number of eggs she lays, based on the needs of the colony. For example, if the colony is experiencing a shortage of food, the queen bee may lay fewer eggs to conserve resources. Similarly, if the population of the colony is dwindling, the queen bee may lay more eggs to increase the population.

Regulating Behavior

In addition to egg laying, the queen bee plays a crucial role in regulating the behavior of the other bees in the colony. The queen bee releases pheromones that alert the other bees to her presence and general well-being. These pheromones help to regulate the behavior of the other bees and ensure that the colony is functioning efficiently.

One of the most important pheromones produced by the queen bee is the queen mandibular pheromone. This pheromone inhibits the development of worker bee ovaries, ensuring that the workers remain sterile and focused on their tasks. By regulating the reproductive behavior of the other bees, the queen bee helps to maintain the division of labor within the colony.

The queen bee also produces other pheromones that signal to the other bees that she is healthy and laying eggs. These pheromones help to regulate the behavior of the other bees and ensure that the colony is functioning efficiently. For example, if the queen bee is injured or sick, the production of pheromones may decrease, which can lead to changes in the behavior of the other bees.

Replacing the Queen Bee

The queen bee has a limited lifespan, and will eventually need to be replaced. When the queen bee begins to decline in health, the worker bees will begin to produce new queen bees.

The process of producing a new queen bee involves the worker bees selecting a few larvae and feeding them royal jelly, a special type of food that is rich in nutrients. The larvae that are fed royal jelly will develop into queen bees, while the other larvae will develop into worker bees. Once the new queen bee emerges from her cell, she will kill any other queen bees that may be present in the colony. The new queen bee will then take over the role of laying eggs and regulating the behavior of the other bees.

Swarming

Another important role of the queen bee is to regulate the swarming behavior of the colony. Swarming is a natural behavior exhibited by bees when a colony becomes too large. During a swarm, a large group of bees will leave the hive and form a new colony elsewhere.

The queen bee plays a crucial role in regulating the swarming behavior of the colony. If the queen bee is healthy and producing pheromones, the swarm is more likely to be successful. If the queen bee is unhealthy or producing fewer pheromones, the swarm is less likely to be successful.

In conclusion, the queen bee plays a vital role in the life of the honey bee colony. The queen bee is responsible for laying eggs, producing pheromones, regulating behavior, and ensuring the health and survival of the colony.

CHAPTER
III
Getting Started in Beekeeping

The importance of location and climate for your hive

The location and climate of a beekeeping can have a substantial effect on a bee colony's productivity and success. Bees require a specific set of environmental conditions to thrive, including access to food sources, suitable temperatures, and protection from predators and other threats. In this section, we'll explore the importance of location and climate in beekeeping, and discuss how beekeepers can choose the best site for their hives.

Access to Food Sources:

One of the most important factors in the success of a bee colony is access to food sources. Bees require nectar and pollen from flowers to produce honey and feed their young. Beekeepers should choose a location with a diverse range of flowering plants that bloom throughout the growing season. The location should also be free from pesticide and herbicide use, as these chemicals can be harmful to bees and other pollinators.

Temperature and Climate:

Bees are cold-blooded insects and require a stable temperature range to survive. The ideal temperature range for a bee colony is between 93 and 95 degrees Fahrenheit (34 to 35 degrees Celsius). Beekeepers should choose a location with a relatively stable temperature range, without extreme fluctuations that could harm the colony. In addition, beekeepers should choose a location with adequate shelter from wind and rain, as extreme weather conditions can be harmful to the hive.

Sun Exposure:

Bees require a certain amount of sun exposure to regulate their body temperature and stimulate foraging. Beekeepers should choose a location with ample sun exposure, but also provide some shade during the hottest parts of the day. A location that receives morning sun and afternoon shade is ideal.

Protection from Predators:

Bees face a variety of predators, including birds, bears, and other animals that are attracted to the sweet smell of honey. Beekeepers should choose a location with adequate protection from predators, such as fencing or electric bear fencing, to ensure the safety of the hive.

Water Source:

Bees require a source of water to regulate the temperature of the hive and dilute honey. Beekeepers should choose a location with access to a clean and reliable water source, such as a nearby stream or pond. In addition, beekeepers may need to provide a supplemental water source for their bees, especially during periods of drought.

Accessibility:

Beekeepers should choose a location that is easily accessible for routine hive maintenance, such as inspecting the hive, adding or removing frames, and harvesting honey. The location should also be easily accessible for moving the hive, in case of extreme weather or other emergencies.

Local Regulations:

Beekeepers should also be aware of local regulations regarding beekeeping, such as zoning laws, nuisance ordinances, and licensing requirements. It is important to obtain any necessary permits or licenses and to follow all local regulations to ensure the legality and safety of the beekeeping operation.

Choosing the right location and climate for a beekeeping site is crucial for the success and productivity of the hive. By considering factors such as access to food sources, temperature and climate, sun exposure, protection from predators, water source, accessibility, and local regulations, beekeepers can choose the best location for their hives and ensure the health and productivity of their bees.

In addition, beekeepers should continually monitor the environmental conditions of their beekeeping site and make adjustments as necessary. This may include adding shade structures, providing supplemental water sources, or moving the hive to a more suitable location. By staying attuned to the needs of the hive and the environmental conditions of the site, beekeepers can ensure the long-term health and productivity of their colonies.

Understanding the different types of hives and their pros and cons

The type of hive used for beekeeping can have a significant impact on the health and productivity of the bee colony. In this section, we will explore the different types of hives commonly used in beekeeping, as well as their pros and cons.

Langstroth Hive

The Langstroth hive is the most common type of hive used in commercial beekeeping. It consists of stacked boxes with frames that are used to support the honeycomb. The frames are removable, allowing beekeepers to inspect the hive and harvest honey.

Pros:

- Provides ample space for the bees to build comb and store honey

- Easy to manage and maintain

- Provides good ventilation for the bees

- Easy to transport and move

- Allows for easy inspection and management of the hive

Cons:

- Can be expensive to purchase and maintain

- Requires a significant amount of equipment and infrastructure

- Can be heavy and difficult to move when fully loaded with honey

- Can be difficult to manage in cold climates, as the bees may need extra insulation

Top Bar Hive

The top bar hive is a simple and inexpensive type of hive that is commonly used in developing countries. It consists of a long, horizontal box with bars placed across the top to support the honeycomb. The bars are typically made of wood or bamboo.

Pros:

- Simple and inexpensive to construct

- Easy to manage and maintain

- Provides good ventilation for the bees

- Reduces the likelihood of swarming

- Can be used in areas with limited resources or infrastructure

Cons:

- Limited space for the bees to build comb and store honey

- Difficult to move or transport

- Limited access to the honeycomb, which can make it difficult to harvest honey

- Not suitable for cold climates, as the horizontal design can lead to poor insulation

Warre Hive

The Warre hive is a vertical hive design that is similar to the top bar hive, but with added boxes that allow for expansion of the hive. The boxes are added to the bottom of the hive as needed, allowing the bees to build comb and store honey as they need.

Pros:

- Provides ample space for the bees to build comb and store honey

- Easy to manage and maintain

- Provides good ventilation for the bees

- Easy to move and transport

- Reduces the likelihood of swarming

Cons:

- Can be expensive to purchase and maintain

- May require additional equipment for expansion of the hive

- Limited access to the honeycomb, which can make it difficult to harvest honey

- Not suitable for cold climates, as the vertical design can lead to poor insulation

Flow Hive

The Flow hive is a relatively new type of hive that allows for easy harvesting of honey without disturbing the bees. It consists of stacked boxes with frames that are used to support the honeycomb. The frames have a specially designed channel that allows for easy extraction of honey.

Pros:

- Easy to harvest honey without disturbing the bees

- Reduces the amount of equipment needed for harvesting honey

- Easy to manage and maintain

- Provides good ventilation for the bees

Cons:

- Can be expensive to purchase and maintain

- Limited access to the honeycomb, which can make it difficult to inspect and manage the hive

- May not be suitable for commercial beekeeping, as the cost may outweigh the benefits

- Not suitable for cold climates, as the vertical design can lead to poor insulation

Horizontal Hive

The horizontal hive is a low-profile design that is similar to the top bar hive, but with added boxes that allow for expansion of the hive. The boxes are added to the side of the hive as needed, allowing the bees to build comb and store honey as they need.

Pros:

- Provides ample space for the bees to build comb and store honey

- Easy to manage and maintain

- Provides good ventilation for the bees

- Can be used in areas with limited resources or infrastructure

- Reduces the likelihood of swarming

Cons:

- Limited access to the honeycomb, which can make it difficult to inspect and manage the hive

- Can be heavy and difficult to move when fully loaded with honey

- May require additional equipment for expansion of the hive

- Not suitable for cold climates, as the horizontal design can lead to poor insulation

Observation Hive

The observation hive is a small hive designed for educational or display purposes. It consists of a small, clear box with frames that allow for observation of the bees and their behavior.

Pros:

- Provides an up-close view of the bees and their behavior

- Can be used for educational or display purposes

- Can be used indoors or outdoors

- Easy to manage and maintain

Cons:

- Limited space for the bees to build comb and store honey

- Limited access to the honeycomb, which can make it difficult to harvest honey

- Not suitable for commercial beekeeping, as the size is too small for a large colony

- May not provide the ideal living conditions for the bees, as the clear box may allow too much light and heat to enter the hive

In conclusion, there are several different types of hives available for beekeeping, each with its own pros and cons. The choice of hive will depend on a variety of factors, including the goals of the beekeeper, the local climate, and the resources available.

Regardless of the type of hive used, it is important to ensure that the hive provides adequate space for the bees to build comb and store honey, and that it provides good ventilation and insulation to maintain the health of the colony. With proper management and care, any type of hive can be used to maintain a healthy and productive bee colony.

Necessary equipment for beekeeping

Beekeeping is a rewarding and fascinating hobby that requires a certain amount of specialized equipment. The right equipment can make beekeeping more efficient, productive, and safe, while the wrong equipment can lead to problems such as disease, low productivity, or even injury. In this section, we will explore the necessary equipment for beekeeping.

Hive

The hive is the most important piece of equipment for beekeeping. It is the home of the bees and provides a place for them to build their honeycomb and store honey. There are several types of hives available, including the Langstroth hive, top bar hive, Warre hive, and Flow hive. The choice of hive will depend on personal preference, local climate, and other factors.

Frames

Frames are used in hives to support the honeycomb and provide a place for the bees to build their comb. They can be made of wood or plastic and come in a variety of sizes to fit different hive types. Frames can also be used to hold foundation, which is a thin sheet of beeswax or plastic that the bees use as a guide to build their comb.

Protective Clothing

Beekeeping protective clothing is essential for the safety of the beekeeper. Bee stings can be painful and even life-threatening for people with allergies, so it's important to wear protective clothing that covers the entire body. The following items are commonly used in beekeeping:

Bee Suit: A bee suit is a full-body suit that covers the beekeeper from head to toe. The suit is made of a lightweight, breathable fabric that provides protection from bee stings while allowing the beekeeper to move freely. Bee suits are available in a range of sizes and styles, from basic cotton suits to high-tech ventilated suits.

Veil: A veil is a protective piece of clothing that covers the head and face, preventing bees from getting into the beekeeper's eyes, nose, and mouth. Veils can be attached to a bee suit or worn separately. Some veils have a hat or helmet attached for additional protection.

Gloves: Beekeeping gloves are made of a thick, protective material that provides protection from bee stings. Gloves can be made of leather, canvas, or rubber. Some beekeepers prefer gloves that allow them to feel more of the comb and the bees' movements.

Boots: Beekeeping boots are made of a sturdy material that provides protection from bee stings. They should be high enough to prevent bees from getting into the beekeeper's pants. Rubber boots are a popular choice as they can be easily cleaned and disinfected.

Protective clothing is an essential part of beekeeping and should be worn at all times when working with the bees. The suit, veil, gloves, and boots should be kept clean and in good condition to ensure their effectiveness.

It's important to choose protective clothing that fits well and is comfortable to wear. It's also important to ensure that the clothing is well-maintained and free of holes or tears that could allow bees to enter.

Hive Tool

An additional necessary piece of gear for beekeeping is the hive tool. It is used to pry apart frames, scrape excess propolis or wax from the hive, and to remove frames for inspection or harvest.

A typical hive tool is a small, flat metal tool with a pointed end and a curved end. The pointed end is used for prying and scraping, while the curved end is used for lifting and separating frames.

It's important to choose a hive tool that is well-made and durable, as it will be used frequently and can become damaged over time.

Smoker

The smoker is another essential piece of equipment for beekeeping. It is used to calm the bees and make them less likely to sting during inspections or harvests.

A typical smoker consists of a metal canister with a bellows attached to one end. The canister is filled with fuel, such as wood chips or burlap, which is ignited to produce smoke.

It's important to choose a smoker that is well-made and durable, as it will be used frequently and can become damaged over time. It's also important to ensure that the smoker is used safely and that the fuel is extinguished properly after use.

Bee Brush

The bee brush is another important piece of equipment for beekeeping. It is used to gently brush bees off of frames or other surfaces during inspections or harvests.

A typical bee brush consists of a soft-bristled brush with a long handle. The bristles are long and soft, allowing the beekeeper to gently move the bees without harming them.

It's important to choose a bee brush that is well-made and durable, as it will be used frequently and can become damaged over time.

Honey Extractor

A honey extractor is a piece of equipment used to extract honey from the comb. It works by spinning the frames in a drum to remove the honey. Honey extractors come in manual and electric versions and can be made of stainless steel or plastic.

Feeder

The feeder is another important piece of equipment for beekeeping. It is used to provide bees with a supplemental source of food during times of low nectar flow or when they are establishing a new hive.

A typical feeder consists of a container that is filled with sugar syrup or other liquid food. The feeder is placed inside the hive, and the bees are able to access the food through small openings in the container.

It's important to choose a feeder that is well-made and durable, as it will be used frequently and can become damaged over time. It's also important to ensure that the feeder is clean and free of contaminants that could harm the bees.

Queen Excluder

The queen excluder is another important piece of equipment for beekeeping. It is a device used in beekeeping that is designed to restrict the movement of the queen bee in the hive. It is typically a flat, perforated metal or plastic panel that is placed between the brood box and the honey super in a Langstroth hive. The purpose of the queen excluder is to keep the queen from laying eggs in the honey supers, which can make it difficult to harvest honey.

Queen excluders work by allowing worker bees to move freely between the brood box and the honey supers, while preventing the larger queen from passing through the small perforations in the excluder. The excluder is typically placed between the brood box and the honey super during the honey flow season, which is the time when bees are actively collecting nectar and pollen to produce honey.

There are several types of queen excluders available, including metal excluders, plastic excluders, and slatted excluders. Metal excluders are the most common type and are made of thin, flat sheets of metal that are perforated with small holes. Plastic excluders are similar to metal excluders but are made of plastic instead of metal. Slatted excluders are made of wooden slats that are spaced closely together to prevent the queen from passing through.

While queen excluders are useful tools in beekeeping, there are some potential drawbacks to using them. One of the most significant concerns is that queen excluders can restrict the movement of the bees in the hive, which can lead to congestion and overcrowding. This can be especially problematic during the winter months, when the bees are less active and need to cluster together to stay warm.

Another potential issue with queen excluders is that they can be difficult to clean and maintain. Over time, propolis and wax can accumulate on the excluder, which can make it less effective at preventing the queen from passing through. Regular cleaning and maintenance are necessary to ensure that the excluder is working properly.

Despite these potential issues, queen excluders are still widely used in beekeeping and are an important tool for beekeepers who want to maximize honey production while minimizing the risk of the queen laying eggs in the honey supers. By providing a barrier between the brood box and the honey supers, queen excluders help to ensure that the bees can produce honey efficiently and that the honey is of high quality.

There are also other tools and accessories that can be helpful in beekeeping. Here are some additional items to consider:

Queen Cage

A queen cage is a small cage used to contain the queen bee during hive inspections or when introducing a new queen to the hive. The cage has a few small holes for ventilation and can be made of plastic or metal. Queen cages can be purchased from beekeeping suppliers or made at home.

Queen Marking Kit

A queen marking kit is a tool used to mark the queen bee with a colored dot to make it easier to identify her during hive inspections. The kit typically includes a marking pen and a catch cage to hold the queen during the marking process.

Bee Feed

Bee feed is used to supplement the bees' diet during times when nectar and pollen are scarce. Bee feed typically consists of a sugar-water mixture and can be provided in a variety of ways, including feeders and pails. Bee feed is particularly important during the winter months when the bees are less active and may not have access to natural food sources.

Bee Escape

A bee escape is a device used to remove bees from honey supers before harvesting the honey. The bee escape is placed in the hive between the honey supers and the brood box, and it allows the bees to move down into the brood box but prevents them from returning to the honey supers.

Hive Stand

A hive stand is a platform used to elevate the hive off the ground. Elevating the hive can provide better ventilation and can help to protect the hive from moisture and pests. Hive stands can be made of wood, metal, or other materials and come in a variety of sizes.

Beekeeping Tools and Accessories

There are many other tools and accessories that can be useful in beekeeping, including pollen traps, honey gates, uncapping knives, and beekeeping journals. Beekeeping suppliers offer a wide range of tools and accessories to help beekeepers manage their hives and harvest honey.

In conclusion, beekeeping requires some essential equipment to properly care for the bees and maintain a healthy hive. The hive, frames, smoker, protective clothing, hive tool, bee brush, honey extractor, feeder, and queen excluder are all necessary for beekeeping. There are also other tools and accessories that can be helpful in beekeeping, such as queen cages, queen marking kits, bee feed, bee escapes, hive stands, and other tools and accessories. By investing in the right equipment and tools, beekeepers can successfully manage their hives and enjoy the many benefits of beekeeping.

Assembling and painting your hive

Assembling a Hive

Assembling a hive for beekeeping is a task that requires careful attention to detail and patience. The hive is the home for the bees and must be constructed in a way that is comfortable and safe for the bees. In this section, we will discuss the steps involved in assembling your hive.

1. Choose the Right Hive

The first step in assembling your hive is to choose the right hive for your needs. There are several types of hives available, including Langstroth hives, Top Bar hives, and Warre hives. Each type of hive has its own advantages and disadvantages, so it's important to do your research and choose the right hive for your needs.

2. Gather Materials and Tools

Once you have chosen the right hive, you will need to gather all of the necessary materials and tools. This includes the hive components (bottom board, brood box, honey supers, inner cover, and outer cover), frames, foundation, nails or screws, a hammer or screwdriver, and a hive tool.

3. Construct the Bottom Board

The bottom board is the foundation of the hive and provides a stable base for the hive. Begin by assembling the sides of the bottom board using nails or screws. Make sure the corners are square and the sides are flush. Then, attach the bottom board to the sides.

4. Build the Brood Box

The brood box is the first box that is placed on top of the bottom board and is where the queen bee lays her eggs. Assemble the sides of the brood box using nails or screws, making sure the corners are square and the sides are flush. Insert the frames and foundation into the brood box.

5. Add Honey Supers

Honey supers are boxes that are placed on top of the brood box and are used to store honey. Assemble the honey supers in the same way as the brood box, making sure the corners are square and the sides are flush. Insert the frames and foundation into the honey supers.

6. Add the Inner Cover and Outer Cover

The inner cover is placed on top of the honey supers and provides ventilation for the hive. The outer cover is placed on top of the inner cover and protects the hive from the elements. Assemble the inner and outer covers according to the manufacturer's instructions.

7. Use the Hive Tool to Insert the Frames

Once all of the hive components are assembled, use the hive tool to insert the frames into the brood box and honey supers. The frames should be spaced evenly and should be in contact with each other to encourage the bees to build straight comb.

8. Place the Hive in a Suitable Location

After the hive is assembled, it should be placed in a suitable location. The hive should be placed in an area that receives plenty of sunlight, is sheltered from the wind, and is away from areas where people or animals may disturb the hive.

Painting your Hive

Painting your hive is an important step in beekeeping that not only enhances the appearance of the hive but also helps to protect the wood from the elements and pests. A well-painted hive can last for many years and provide a comfortable and healthy home for your bees. Here are the steps involved in painting your hive.

1. Choose the Right Paint

The first step in painting your hive is to choose the right paint. It's important to use paint that is safe for bees and won't harm them. Avoid using oil-based paints, as they can be toxic to bees. Instead, choose a high-quality, water-based paint or stain that is safe for use around bees.

2. Prepare the Surface

Before you begin painting, it's important to prepare the surface of the hive. Use a scraper or sandpaper to remove any rough spots, splinters, or old paint. This will help the paint adhere to the wood and give the hive a smooth finish.

3. Apply Primer

Once the surface is prepared, apply a coat of primer to the hive. The primer will help the paint adhere to the wood and provide an extra layer of protection. Be sure to cover all of the surfaces of the hive, including the inside and bottom of the boxes.

4. Apply Paint

Apply a coat of paint or stain to the hive once the priming has dried. Be sure to cover all of the surfaces of the hive, including the inside and bottom of the boxes. Apply a second coat of paint or stain if necessary, allowing the first coat to dry completely before applying the second coat.

5. Allow to Dry

After the final coat of paint is applied, allow the paint to dry completely. Depending on the type of paint and the weather conditions, this may take several days to several weeks. It's important to allow the paint to cure completely before adding bees to the hive.

6. Add Finishing Touches

Once the paint is dry, you can add any finishing touches to the hive, such as decorative accents or stenciling. Be sure to use non-toxic materials and avoid applying anything that may be harmful to bees.

7. Maintain the Paint

It's important to maintain the paint on your hive to ensure it lasts for many years. Check the paint regularly for any signs of peeling or cracking and touch up any areas that need it. Regular maintenance will help to protect the hive from the elements and pests and ensure a long life for your hive.

In conclusion, assembling and painting your hive is an important part of beekeeping. A well-constructed and painted hive can provide a comfortable and healthy home for your bees and help to protect the hive from the elements and pests. By following the steps outlined in this section, you can assemble and paint your hive with confidence and enjoy the many benefits of beekeeping.

Acquiring bees and introducing them to the hive

Acquiring bees and introducing them to a hive can be a daunting task, especially for beginners. In this article, we will provide a comprehensive guide on acquiring bees and introducing them to a hive, covering everything from choosing a type of bee to ongoing care and maintenance.

Choosing a Type of Bee:

The first step in acquiring bees is choosing a type of bee. For beekeeping, the most commonly kept bees are honeybees. Within honeybees, there are different breeds that are well-suited for beekeeping. Here are some of the most popular types of bees for beekeeping:

1. Italian Honeybees: Italian honeybees are the most commonly kept bees in the United States. They are known for their gentle temperament and their capacity to make a lot of honey. Italian bees are also good at foraging and are resistant to some diseases, such as American Foulbrood.

2. Carniolan Honeybees: Carniolan honeybees are native to the Alps region and are known for their hardiness and ability to survive in cold climates. They are also good at building comb and brood rearing. Carniolan bees are gentle and have a low tendency to swarm.

3. Russian Honeybees: Russian honeybees are a subspecies of the Western honeybee and are known for their resistance to Varroa mites, which are a common pest in honeybee colonies. Russian bees are also good at foraging and are less prone to robbing (stealing honey from other hives).

4. Buckfast Honeybees: Buckfast honeybees are a hybrid breed created by a monk named Brother Adam in the 20th century. They are known for their gentleness, productivity, and resistance to disease. Buckfast bees are also good at foraging and are adaptable to different climates.

Each breed has its own characteristics, such as temperament, honey production, and disease resistance. It's important to do research and choose a breed that is well-suited for your climate and environment.

When choosing a type of bee, it's also important to consider whether you want a package or a nuc (nucleus colony). A package is a box that contains a certain number of bees and a queen, while a nuc is a small colony that includes frames of brood, honey, and pollen in addition to bees and a queen. Nucs are more expensive than packages, but they offer several advantages, such as a head start in colony development and better acclimation to the hive.

Finding a Reputable Supplier:

Once you have decided on the type of bee you want, it's important to find a reputable supplier. You can search online for bee suppliers in your area, or ask local beekeeping associations for recommendations. It's important to choose a supplier that has a good reputation and is known for providing healthy, disease-free bees. You should also ask about the supplier's breeding and selection practices, as well as any guarantees or warranties they offer.

Ordering Bees:

Once you have found a reputable supplier, you can order your bees. Bees are usually sold in packages or nucs, and the number of bees you need will depend on the size of your hive. A standard hive can hold up to 50,000 bees, but beginners may want to start with a smaller colony of 10,000 to 20,000 bees.

When ordering bees, it's important to choose a delivery date that is appropriate for your climate and environment. Bees should be installed in the hive when temperatures are above 50 degrees Fahrenheit and there is plenty of nectar and pollen available. It's also important to ensure that you are available to receive the bees on the delivery date, as they need to be installed in the hive as soon as possible.

Transporting Bees:

Transporting bees can be stressful for the bees, so it's important to take preventative measures to ensure their safety. Bees should be transported in a well-ventilated, dark container that protects them from extreme temperatures and jostling. The container should be marked with the words "Live Bees" and handled carefully during transport.

Cost:

The cost of purchasing bees can vary depending on the type of bee, package vs. nucleus colony, and source of bees. It's important to consider the long-term benefits of investing in healthy, high-quality bees that will establish a productive colony.

Preparing the Hive:

Before the bees arrive, you need to prepare the hive. Make sure the hive is clean, dry, and free of pests and diseases. You also need to add frames with foundation (wax or plastic sheets) for the bees to build comb on. The number of frames you need will depend on the size of your hive and the number of bees you have ordered.

It's also a good idea to add a feeder to the hive, which will provide the bees with a sugar syrup mixture to help them build up their strength and resources. The feeder should be placed near the brood nest, where the bees will have easy access to it.

Installing the Bees:

When your bees arrive, they will usually come in a package or nuc. Follow the instructions provided by your supplier to install the bees in the hive. This usually involves shaking the bees into the hive and releasing the queen. It's important to handle the bees gently and calmly, as they can become agitated and sting if they feel threatened.

After installing the bees, it's important to close up the hive and let the bees settle in for a few days before opening the hive again. During this time, the bees will begin to build comb and establish their colony.

Feeding the Bees:

When you first install the bees, they may not have enough food to sustain themselves. It's important to provide them with a sugar syrup mixture to help them build up their strength and resources. The sugar syrup can be made by mixing one part sugar with one part water and heating the mixture until the sugar dissolves. The mixture should be cooled before adding it to the feeder.

You may need to refill the feeder several times in the first few weeks after installing the bees, as they will be busy building comb and expanding their colony.

Monitoring the Hive:

After installing the bees, you need to monitor the hive regularly to ensure that the bees are healthy and thriving. Check for signs of disease, pests, and swarming. You may also need to add more frames or supers (boxes) to the hive as the colony grows.

It's important to perform regular inspections of the hive, at least once a week during the spring and summer months, and less frequently during the winter. During inspections, you should look for signs of brood (developing bees), honey, and pollen. You should also check for signs of pests, such as mites or beetles, and treat the hive if necessary.

Providing Ongoing Care:

Beekeeping requires ongoing care and maintenance. You will need to monitor the hive regularly, manage pests and diseases, and provide supplemental feeding if needed. You may also need to harvest honey and other bee products.

Harvesting honey can be a complicated process, and it's important to do it at the right time to ensure that the bees have enough honey to survive the winter. The process involves removing frames of honey from the hive and extracting the honey from the comb using a honey extractor.

Acquiring bees and introducing them to a hive can be a rewarding and fulfilling experience, but it requires time, effort, and knowledge to be successful. Choosing the right type of bee, finding a reputable supplier, and preparing the hive are all important steps in the process. Feeding and monitoring the bees, as well as providing ongoing care and maintenance, are essential to the health and well-being of the colony. With proper care and attention, beekeeping can be a fascinating and enjoyable hobby or business.

CHAPTER
IV
Hive Management

Inspecting your Hive

Inspecting your hive is an important task for any beekeeper, as it allows you to monitor the health and productivity of your colony. Regular inspections can help you identify and address any issues or problems, such as pests, diseases, or insufficient food stores. In this article, we will provide a comprehensive guide on inspecting your hive, covering everything from when to inspect to what to look for.

When to Inspect Your Hive:

The frequency of hive inspections will depend on the time of year and the size and strength of your colony. During the spring and summer months, when the colony is actively building up and producing honey, you should inspect the hive at least once a week. During the fall and winter months, when the colony is preparing for winter, you should inspect the hive less frequently, about once every three to four weeks.

It's important to choose a warm, sunny day to inspect your hive, as this will ensure that the bees are active and foraging. Avoid inspecting the hive on a cold or rainy day, as this can stress the bees and disrupt the hive environment.

Preparing for Inspection:

Before inspecting your hive, it's important to prepare yourself and your equipment. To avoid getting stung by bees, put on protective equipment such a bee suit, veil, and gloves. You should also use a smoker to calm the bees and make them less agitated during the inspection.

Make sure that you have all of the necessary equipment on hand, such as a hive tool, brush, and smoker. It's also a good idea to have a notebook or record-keeping system to document your observations and any actions taken.

During the Inspection:

When you begin the inspection, start by observing the bees from the outside of the hive. Look for signs of activity, such as bees entering and exiting the hive, and listen for the buzzing sound of the bees. This will give you an idea of how active and healthy the colony is.

Next, use your hive tool to gently pry apart the frames and inspect the comb. Look for signs of brood (developing bees), honey, and pollen. The brood should be healthy and well-formed, with no signs of disease or pests. The honey should be free of mold or fermentation, and the pollen should be varied and colorful.

While inspecting the comb, you should also look for signs of pests or diseases, such as Varroa mites, wax moths, or foulbrood. If you see any signs of pests or diseases, take appropriate actions to treat or prevent them.

After inspecting the comb, gently brush off any bees that have landed on your clothing or equipment. Avoid crushing or injuring the bees, as this can agitate the colony.

Documenting Your Observations:

After the inspection, it's important to document your observations and any actions taken. This will help you track the health and progress of the colony over time and make informed decisions about future management.

Record the date of the inspection, the weather conditions, and any observations or notes about the health and activity of the colony. If you took any actions, such as treating for pests or diseases, document these actions and the results.

Maintaining Your Hive:

Regular inspections are an important part of maintaining your hive, but there are other tasks that should be done on a regular basis to ensure the health and productivity of the colony. These tasks may include:

- Feeding the bees during times of low nectar flow, such as during droughts or early spring.

- Providing a water source for the bees, such as a shallow dish or bird bath.

- Replacing old or damaged comb to ensure that the hive remains clean and healthy.

- Monitoring the colony for signs of swarming, and taking appropriate actions to prevent or manage swarming.

- Harvesting honey and other bee products at the appropriate times to ensure that the bees have enough stores to survive the winter.

It's important to keep track of the progress of your hive over time and make adjustments as needed to ensure its health and productivity. By regularly inspecting your hive and keeping detailed records, you can monitor the progress of your colony and make informed decisions about future management.

What to Look for During an Inspection?

During an inspection, there are several things to look for to ensure the health and productivity of your hive. Here are some of the most important things to check:

1. Brood: Look for signs of healthy brood, including larvae, pupae, and capped cells. The brood should be evenly distributed throughout the hive, with no signs of disease or pests.

2. Honey and Pollen: Check the honey and pollen stores to ensure that the bees have enough food to sustain themselves. The honey should be free of mold or fermentation, and the pollen should be varied and colorful.

3. Pests and Diseases: Check for indications of illnesses and pests like Varroa mites, wax moths, or foulbrood. If you see any signs of pests or diseases, take appropriate actions to treat or prevent them.

4. Comb: Inspect the comb for signs of damage or decay. Replace old or damaged comb to ensure that the hive remains clean and healthy.

5. Queen: Check for the presence and activity of the queen. Look for eggs, larvae, and pupae, which indicate that the queen is laying eggs and the colony is healthy.

6. Temperament: Observe the behavior and temperament of the bees. Look for signs of aggression or agitation, which can indicate that the hive is stressed or unhealthy.

In conclusion, inspecting your hive is an essential part of beekeeping, as it allows you to monitor the health and productivity of your colony. By inspecting your hive regularly and keeping detailed records, you can ensure a healthy and productive colony that will provide you with honey and other bee products for years to come.

Maintaining Hive Health

Maintaining the health of your beehive is essential for the survival and productivity of your colony. A healthy hive will produce more honey and be more resistant to pests and diseases. In this article, we will provide a comprehensive guide on maintaining hive health, covering everything from feeding and monitoring to disease prevention and treatment.

Feeding Your Bees:

Feeding your bees is an important part of maintaining their health, especially during times of low nectar flow, such as during droughts or early spring. There are several types of feed that you can provide for your bees, including sugar syrup, pollen patties, and fondant.

Sugar syrup is a common feed for bees, especially during times of low nectar flow. To make sugar syrup, mix one part sugar with one part water and bring to a boil, stirring until the sugar is dissolved. Let the syrup cool before feeding it to the bees.

Pollen patties are another type of feed that you can provide for your bees. Pollen patties are made from a mixture of pollen and sugar, and they provide the bees with essential nutrients for brood rearing and honey production.

Fondant is a type of candy that you can provide for your bees during the winter months when there is little or no nectar available. Fondant is made from sugar, water, and corn syrup, and it provides the bees with a source of food that they can eat during the winter.

Monitoring Your Hive:

Monitoring your hive is an important part of maintaining its health. Regular inspections can help you identify and address any issues or problems, such as pests, diseases, or insufficient food stores. During an inspection, look for signs of brood, honey, and pollen, as well as any signs of pests or diseases.

You should also monitor the activity of the bees outside the hive. Look for signs of activity, such as bees entering and exiting the hive, and listen for the buzzing sound of the bees. This will give you an idea of how active and healthy the colony is.

Pest and Disease Prevention:

Preventing pests and diseases is essential for maintaining the health of your hive. These are some recommendations for avoiding diseases and pests:

1. Keep your hive clean: Regularly clean your hive and remove any debris or dead bees. This will help prevent the buildup of pests and diseases.

2. Use screened bottom boards: Screened bottom boards can help prevent the buildup of mites and other pests in the hive.

3. Use natural treatments: There are several natural treatments that you can use to prevent pests and diseases, such as essential oils and diatomaceous earth.

4. Monitor for pests and diseases: Regularly monitor your hive for signs of pests and diseases, such as mites or foulbrood. The prevention of the spread of illnesses and pests depends on early detection.

Disease Treatment:

If you do encounter a pest or disease in your hive, it's important to take appropriate actions to treat it. Here are some tips for treating pests and diseases:

1. Use natural treatments: There are several natural treatments that you can use to treat pests and diseases, such as essential oils and organic acids.

2. Follow label instructions: When using chemical treatments, make sure to follow the label instructions carefully. Improper use can be harmful to your bees and the environment.

3. Remove infected comb: If you encounter a disease such as foulbrood, it's important to remove and dispose of infected comb to prevent the spread of the disease.

4. Requeen: In some cases, it may be necessary to requeen the hive to prevent the spread of disease.

Preventing and treating pests and diseases is also important for maintaining the health of your hive, as it can prevent the spread of harmful agents that can damage your colony.

In addition to feeding, monitoring, and pest and disease prevention, there are other tasks that you can do to maintain the health of your hive. These tasks include:

1. Replacing old or damaged comb: Old or damaged comb can harbor pests and diseases, so it's important to replace it with new comb when necessary.

2. Providing a water source: Bees need water to produce honey and to cool the hive, so providing a water source near the hive is important.

3. Controlling swarming: Swarming is a natural process, but it can reduce the productivity of your hive. To control swarming, you can split your hive or provide additional space for the bees to expand.

4. Harvesting honey: Harvesting honey is an important part of beekeeping, but it's important to do it at the appropriate time and in the correct manner to avoid stressing the bees and damaging the hive.

5. Maintaining a clean and organized hive: Keeping your hive clean and organized can help prevent the buildup of pests and diseases, and it can also make inspections and maintenance tasks easier.

In summary, by feeding your bees, monitoring their activity, preventing and treating pests and diseases, and performing other maintenance tasks, you can ensure a healthy and productive hive.

Recognizing and preventing common bee diseases and pests

Bees are susceptible to various diseases and pests, which can damage or destroy the hive, weaken the health of the bees, and reduce honey production. As a beekeeper, it is important to recognize the signs and symptoms of common bee diseases and pests and take appropriate measures to prevent or control them. In this article, we will provide a detailed guide on recognizing and preventing common bee diseases and pests.

Common Bee Diseases:

1. American Foulbrood:

American Foulbrood (AFB) is a highly contagious bacterial disease that affects the brood of the hive. AFB can cause the death of entire colonies if left untreated. Symptoms of AFB include sunken, perforated, or discolored brood cells, and a foul odor. The spores of the bacteria can remain viable for many years, so it is important to practice good hygiene and sanitation practices in the hive.

Prevention: To prevent the spread of AFB, it is important to practice good hygiene and sanitation practices in the hive. Replace old or damaged comb, and avoid feeding your bees honey that you suspect may be contaminated. Additionally, be sure to inspect new colonies for signs of AFB before introducing them to your apiary.

Treatment: AFB is a serious disease, and infected colonies should be destroyed to prevent the spread of the disease. Burn or bury the infected hive and frames, and sterilize all equipment that has come into contact with the infected hive.

2. European Foulbrood:

European Foulbrood (EFB) is a bacterial disease that affects the brood of the hive. It is less severe than AFB but can still weaken the colony. Symptoms of EFB include irregular and sunken brood cells, and a yellow or brown color.

Prevention: To prevent the spread of EFB, practice good hygiene and sanitation practices in the hive. Replace old or damaged comb, and avoid feeding your bees honey that you suspect may be contaminated.

Treatment: EFB can be treated with antibiotics, but it is important to consult with a veterinarian or local beekeeping organization before administering any medications.

3. Nosema:

Nosema is a fungal disease that affects the digestive system of the bees. It can cause dysentery and reduce the lifespan of the bees. Symptoms of nosema include diarrhea and weakened bees.

Prevention: To prevent the spread of nosema, provide your bees with clean water and avoid feeding them old or fermented pollen. You can also use natural treatments, such as essential oils, to prevent the spread of nosema.

Treatment: Nosema can be treated with medication, but it is important to consult with a veterinarian or local beekeeping organization before administering any medications.

Common Bee Pests:

1. Varroa Mites:

Varroa mites are a common pest that can weaken and kill colonies. They attach themselves to the bees and feed on their blood, weakening them and spreading diseases. Symptoms of Varroa mite infestations include deformed wings, crawling bees, and low honey production.

Prevention: To prevent Varroa mite infestations, use screened bottom boards to prevent mites from entering the hive. You can also use natural treatments, such as essential oils, to control Varroa mites.

Treatment: Varroa mites can be treated with chemical or natural treatments. It is important to follow label instructions carefully when using chemical treatments.

2. Wax Moths:

Wax moths are another common pest that can damage the comb and weaken the colony. They lay their eggs on the comb, and the larvae feed on the wax and other debris. Symptoms of wax moth infestations include webbing and moth larvae on the comb.

Prevention: To prevent wax moth infestations, maintain a clean and organized hive. Remove old or damaged comb, and avoid storing comb in humid or warm areas.

Treatment: Wax moth infestations can be controlled by freezing the infested comb or exposing it to sunlight for several hours. You can also use natural treatments, such as essential oils, to prevent the spread of wax moths.

3. *Small Hive Beetle:*

Small hive beetles are a pest that can damage the comb and weaken the colony. They lay their eggs on the comb, and the larvae feed on the honey and pollen. Symptoms of small hive beetle infestations include damaged comb and beetles crawling on the comb.

Prevention: To prevent small hive beetle infestations, maintain a clean and organized hive. Remove old or damaged comb, and use beetle traps or natural treatments to control the beetles.

Treatment: Small hive beetle infestations can be treated with natural treatments, such as diatomaceous earth or essential oils. You can also use chemical treatments, but it is important to follow label instructions carefully.

Other pests to be aware of include ants, mice, and birds. These pests can damage the hive and steal honey, and they should be prevented or controlled as needed.

In conclusion, recognizing and preventing common bee diseases and pests is essential for maintaining the health and productivity of your hive. By practicing good hygiene and sanitation practices, using natural treatments, and monitoring the activity of your bees, you can prevent the spread of harmful agents that can damage your colony.

Honey harvesting and extraction

Honey harvesting and extraction is an important aspect of beekeeping, as it allows beekeepers to harvest the honey produced by their bees. Honey is a natural sweetener that is rich in vitamins, minerals, and antioxidants, making it a valuable product for both personal use and commercial sale. In this section, we will provide a detailed guide on honey harvesting and extraction.

<u>When to Harvest Honey:</u>

The timing of honey harvesting depends on several factors, including the location, weather, and strength of the hive. In general, honey can be harvested when the majority of the cells in the honeycomb are capped, indicating that the honey is fully ripened. However, it is important to leave enough honey for the bees to survive the winter.

Before harvesting honey, it is important to inspect the hive and make sure that the bees are healthy and active. Avoid harvesting honey during periods of drought or other environmental stresses, as this can weaken the colony and reduce the quality of the honey.

Preparing for Honey Harvesting:

To prepare for honey harvesting, you will need the following equipment:

- Protective clothing: Wear a full bee suit, gloves, and a veil to protect yourself from bee stings.

- Smoker: Use a smoker to calm the bees and prevent them from becoming agitated.

- Honey supers: These are boxes that fit on top of the brood box and hold the frames of honey.

- Bee brush: Use a bee brush to gently brush the bees off the frames of honey.

- Uncapping knife or tool: Use an uncapping knife or tool to remove the wax caps from the honeycomb.

- Honey extractor: A honey extractor is a machine that extracts honey from the comb.

<u>Harvesting Honey:</u>

To harvest honey, follow these steps:

1. Use the smoker to calm the bees and prevent them from becoming agitated.

2. Remove the honey supers from the hive and brush the bees off the frames of honey.

3. Place the frames of honey in a covered container to transport them to the honey extraction area.

4. In the honey extraction area, use an uncapping knife or tool to remove the wax caps from the honeycomb.

5. Place the uncapped frames of honey in the honey extractor, and spin the frames to extract the honey.

6. Once the honey is extracted, filter it through a strainer or cheesecloth to remove any wax or debris.

7. Store the honey in clean, dry containers.

Cleaning Up:

After harvesting honey, it is important to clean up the equipment and the hive. Be sure to remove any wax or debris from the honey extractor and other equipment, and store them in a dry, cool place. Return the honey supers to the hive, and allow the bees to clean up any remaining honey.

Tips for Successful Honey Harvesting:

1. Avoid harvesting honey during periods of drought or other environmental stresses.

2. Inspect the hive before harvesting honey to ensure that the bees are healthy and active.

3. Use protective clothing and a smoker to prevent bee stings and calm the bees.

4. Use a bee brush to gently remove bees from the frames of honey.

5. To take the wax caps off the honeycomb, use an uncapping tool or knife.

6. Use a honey extractor to extract the honey from the comb.

7. Filter the honey through a strainer or cheesecloth to remove any wax or debris.

8. Store the honey in clean, dry containers.

Honey Extraction:

The honey extraction process involves removing the honey from the honeycomb and separating it from the wax. There are two main methods of honey extraction: uncapping and spinning.

Uncapping: In this method, the wax caps are removed from the honeycomb cells using an uncapping knife or tool. The uncapped frames are then placed in a honey extractor, which spins the frames to release the honey from the cells. The honey then flows out of the extractor and can be collected in a container.

Spinning: In this method, the frames of honey are placed in a centrifuge, which spins the frames to extract the honey. The honey then flows out of the extractor and can be collected in a container.

Regardless of the method used, it is important to filter the honey to remove any wax or debris. This can be done by using a strainer or cheesecloth.

Storing Honey:

Once the honey has been extracted and filtered, it must be properly stored to preserve its quality and freshness. Honey should be stored in clean, dry containers that are tightly sealed to prevent moisture from entering. Glass jars with screw-top lids are a popular choice for storing honey.

Honey can crystallize over time, which is a natural process that does not affect its quality. To liquefy crystallized honey, place the jar in a warm water bath or microwave it for short intervals at low power.

Honey harvesting and extraction is a rewarding aspect of beekeeping, but it requires careful preparation and attention to detail. By following proper techniques and using the appropriate equipment, you can ensure the safety and health of your bees, and produce high-quality honey. With proper care and attention, beekeeping can be a fascinating and rewarding hobby or business, and can provide a valuable source of natural honey for personal use or commercial sale.

Preparing the hive for winter

Preparing a beehive for winter is crucial for the survival of the colony. Being cold-blooded, bees depend on their surroundings to maintain a consistent body temperature. It can be deadly for bees when temperatures fall below freezing throughout the winter. In this section, we will discuss the steps necessary for preparing the hive for winter, ensuring the survival of the colony.

1. Assess the health of the colony

Before winterizing the hive, it's important to assess the health of the colony. A weak or sickly colony will have a much harder time surviving the winter than a healthy one. Look for signs of disease, pests, or other issues that could harm the colony. If you notice any problems, address them before winter arrives.

2. Provide adequate food stores

Bees require a significant amount of food to survive the winter. As the temperature drops, the bees will huddle together in the hive to stay warm. This procedure consumes a lot of energy, so that the bees will need a large supply of honey to sustain them through the winter.

It's important to make sure the hive has enough food stores before the cold weather sets in.

To determine if the colony has enough food, you can conduct a simple inspection. Lift the top box of the hive and check the weight. If it feels light, the bees may not have enough food to survive the winter. If this is the case, you can add more honey frames or sugar syrup to the hive.

3. Reduce the entrance size

During the winter, the hive needs to be sealed to keep out cold drafts. One way to do this is to reduce the size of the entrance. A smaller entrance will make it easier for the bees to defend the hive against predators while also reducing the amount of cold air that can enter the hive.

You can reduce the entrance size by using a wooden or metal entrance reducer. Simply slide the reducer into the entrance of the hive to make it smaller. You can also use a piece of foam or weather stripping to seal the entrance.

4. Provide insulation

Insulating the hive is crucial for keeping the bees warm during the winter. There are several ways to insulate a hive, including wrapping it with insulation material or placing insulation boards on the top of the hive.

If you choose to wrap the hive, you can use materials like burlap, tar paper, or foam insulation. Make sure to leave the entrance uncovered and provide ventilation to prevent condensation from building up inside the hive.

Insulation boards can be placed on top of the hive to provide an additional layer of protection. These boards can be made from materials like foam insulation, polystyrene, or straw.

5. Provide ventilation

Although the hive needs to be sealed to keep out the cold, it's also important to provide ventilation to prevent condensation from building up inside the hive. Condensation can be harmful to the bees and can lead to the growth of mold and fungus.

To provide ventilation, you can place a small piece of wood or foam under the outer cover of the hive to create a small gap. This gap will allow moisture to escape while also keeping the hive sealed.

6. Monitor the hive throughout the winter

Even with proper preparation, there is always a risk that the colony will not survive the winter. It's important to monitor the hive throughout the winter to ensure that the bees are healthy and have enough food.

On warmer days, you can quickly lift the outer cover of the hive to check the food stores and assess the health of the bees. If the colony seems weak or is running low on food, you can provide additional food or take other steps to support the colony.

7. Consider supplemental feeding

If you live in an area with long winters or harsh weather, you may need to provide supplemental feeding to help the bees survive. This can involve feeding the bees sugar syrup or pollen patties to supplement their food stores.

Making sugar syrup involves combining sugar and water in a 1:1 ratio. You can place the syrup in a feeder inside the hive, or you can add it directly to the frames.

Pollen patties can be made by mixing pollen and sugar with a small amount of water to form a paste. These patties can be placed on top of the frames inside the hive.

8. Treat for pests and diseases

Winter can be a challenging time for bees, and it's important to make sure the colony is healthy and free from pests and diseases. Before winter sets in, you should treat the hive for any known pests or diseases.

Common pests include mites, wax moths, and hive beetles. There are a variety of treatments available for these pests, including chemical treatments and natural remedies.

Diseases like American Foulbrood and Nosema can also be devastating to a bee colony. If you suspect that your colony has a disease, it's important to seek help from a local beekeeping association or expert.

9. Protect the hive from wind and snow

Strong winds and heavy snow can be damaging to a beehive. If your hive is located in an area that is prone to these weather conditions, you may need to take additional steps to protect the hive.

You can use windbreaks or snow fences to protect the hive from the elements. You can also place a tarp or other protective cover over the hive to prevent snow from accumulating on the roof.

10. Plan for spring

Although winter can be a challenging time for bees, it's also an opportunity to plan for the spring. During the winter, beekeepers can take the time to prepare for the coming season by ordering new equipment, planning for new colonies, and researching new beekeeping techniques.

By taking steps to prepare the hive for winter, beekeepers can help ensure the survival of their colonies. With proper preparation and care, bees can thrive even in the harshest of winter conditions.

CHAPTER
V
Advanced Beekeeping Techniques

Queen Rearing

Queen rearing is the process of producing new queen bees for a beehive. This is an essential part of beekeeping, as the queen is the most important member of the colony. She is responsible for laying eggs, controlling the behavior of the other bees, and ensuring the survival of the colony. In this article, we will discuss the basics of queen rearing, including why it's necessary, the methods used, and the timeline for the process.

<u>Why is queen rearing necessary?</u>

Queen rearing is necessary for several reasons. The most common reason is to replace an old or failing queen. As a queen ages, her egg-laying ability declines, which can lead to a decline in the health of the colony. A new queen can help revitalize the colony and ensure its continued success.

Queen rearing is also necessary for beekeepers who want to expand their apiary or produce new colonies for sale. By rearing their own queens, beekeepers can control the genetics of their colonies and ensure that they have healthy and productive bees.

Methods of Queen Rearing

There are several methods of queen rearing, each with its own advantages and disadvantages. The most common methods are the grafting method and the cell punch method.

<u>Grafting Method</u>

The grafting method is the most popular and widely used method of queen rearing. This method involves removing a small larva from a healthy colony and transferring it to a specially prepared cell cup. The larva is then fed a special diet of royal jelly, which helps it develop into a queen bee.

To graft a larva, beekeepers will need a few special tools, including a grafting needle, a cell cup, and a queen rearing frame. The process involves carefully selecting a larva from a healthy colony and transferring it to the cell cup. Once the larva is in the cup, it is placed in a queen rearing frame and placed in a special queen rearing colony.

Cell Punch Method

The cell punch method is a simpler and less labor-intensive method of queen rearing. This method involves using a special tool to remove a section of capped brood from a healthy colony. The section of brood is then placed in a specially prepared cell cup and allowed to develop into a queen.

The cell punch method is easier and less time-consuming than the grafting method, but it does not produce as many queens. Additionally, the queens produced using this method may not be as high quality as those produced using the grafting method.

Timeline for Queen Rearing

Queen rearing is a time-consuming process that requires patience and attention to detail. It's important to note that the timeline for queen rearing can vary depending on a variety of factors, including the climate, the method used, and the health of the colony.

Tips for Successful Queen Rearing

Queen rearing can be a challenging process, but with the right tools and techniques, it can be successful. Here are a few tips for successful queen rearing:

1. Start with a healthy colony. The success of queen rearing depends on the health of the colony. Make sure that the

colony you are using is healthy and free from pests and diseases.

2. Use the right tools. Queen rearing requires a few specialized tools, including a grafting needle, cell cups, and queen rearing frames. Make sure that you have all the necessary tools before starting the process.

3. Choose the right method. There are several methods of queen rearing, each with its own advantages and disadvantages. Choose a method based on your requirements and prior knowledge.

4. Keep the queen rearing colony well-fed. Queen rearing requires a lot of resources, including pollen and nectar. Make sure that the queen rearing colony has plenty of food to support the development of the queen cells.

5. Monitor the development of the queen cells. It's important to monitor the development of the queen cells to ensure that they are developing properly. Remove any cells that are not developing properly.

6. Be patient. Queen rearing is a time-consuming process that requires patience and attention to detail. Don't rush the process and be prepared to make adjustments along the way.

In conclusion, queen rearing is an important part of beekeeping. By rearing their own queens, beekeepers can control the genetics of their colonies and ensure that they have healthy and productive bees. With the right tools and techniques, queen rearing can be a successful and rewarding process.

Bee breeding and genetics

<u>Why is Bee Breeding and Genetics Important?</u>

Bee breeding and genetics are important for several reasons. By selecting bees with desirable traits, beekeepers can improve the health and productivity of their colonies. For example, breeding bees for increased honey production can lead to higher yields and more profit for the beekeeper.

Breeding for disease resistance can help protect the colony from common pests and diseases, reducing the need for chemical treatments. Selecting bees for docility can make it easier and safer for beekeepers to work with the colony.

Breeding and genetics also play a critical role in the conservation of bee populations. Many bee populations around the world are in decline due to factors like habitat loss, pesticide use, and disease. Breeding bees for desirable traits can help improve the health and resilience of bee populations, ensuring their survival for future generations.

<u>Traits that Beekeepers Look for in Bees</u>

Beekeepers look for a variety of traits when breeding bees, including:

1. Honey production: Bees that produce large quantities of honey are desirable for commercial beekeeping operations.

2. Disease resistance: Bees that are resistant to pests and diseases are more likely to survive and thrive.

3. Temperament: Bees that are docile and easy to work with are desirable for safety and ease of handling.

4. Winter hardiness: Bees that can survive harsh winter conditions are more likely to survive and thrive.

5. Pollination ability: Bees that are effective pollinators can help increase crop yields.

6. Hygienic behavior: Bees that exhibit hygienic behavior, such as removing dead or diseased larvae, can help prevent the spread of disease in the colony.

Methods of Bee Breeding

There are several methods of bee breeding, each with its own advantages and disadvantages. The most common methods are selective breeding, artificial insemination, and genetic engineering.

Selective Breeding

Selective breeding is the most common method of bee breeding. This method involves selecting bees with desirable traits and breeding them together to produce offspring with those traits.

The process of selective breeding can be time-consuming, as it requires careful observation and selection of individual bees. However, it can be a cost-effective way to improve the health and productivity of the colony.

Artificial Insemination

Artificial insemination is a more advanced method of bee breeding. This method involves collecting semen from a male bee and using it to inseminate a queen bee. By selecting the best male bees for semen collection, beekeepers can ensure that the offspring will have desirable traits.

Artificial insemination requires specialized equipment and training, and it can be expensive. However, it can be an effective way to rapidly improve the genetics of the colony.

Genetic Engineering

Genetic engineering is a relatively new method of bee breeding. This method involves altering the DNA of bees to produce desired traits, such as disease resistance or increased honey production.

Genetic engineering is a controversial topic, as it raises ethical and environmental concerns. However, it has the potential to produce bees with desirable traits more quickly and efficiently than traditional breeding methods.

Splitting hives and creating new colonies

Splitting hives and creating new colonies is an essential part of beekeeping. It allows beekeepers to expand their apiary, prevent swarming, and maintain the health of their colonies. In this article, we will discuss the basics of splitting hives, including when and why to split, the methods used, and tips for success.

<u>Why Split Hives?</u>

Splitting hives is a common practice in beekeeping. It involves separating a healthy colony into two or more smaller colonies, each with its own queen bee. There are several reasons why beekeepers split hives:

1. To prevent swarming: Swarming is a common bee behavior, although it can be disruptive and damaging to the colony. By splitting the hive, beekeepers can reduce the likelihood of swarming.

2. To create new colonies: Splitting a hive is an easy way to create new colonies. Each split can be turned into a new colony, allowing beekeepers to expand their apiary.

3. To manage disease: If a colony is infected with a disease, splitting the hive can help prevent the disease from spreading to other colonies.

4. To maintain hive health: Splitting the hive can help prevent overcrowding and maintain the health of the colony.

<u>When to Split Hives?</u>

The best time to split a hive is in the spring, when the colony is starting to build up its population. Splitting too early or too late in the season can be detrimental to the health of the colony.

Before splitting a hive, beekeepers should ensure that the colony is healthy and has enough resources to support the new colonies. It's also important to make sure that each new colony has a queen bee and enough resources to support its growth.

Methods of Splitting Hives

There are several methods of splitting hives, each with its own advantages and disadvantages. The most common methods are the two-box method, the nucleus method, and the walk-away split method.

1. Two-box method: The two-box method involves splitting the hive into two equal parts, each with its own brood and honey stores. The queen bee is placed in one of the boxes, and the other box is left to raise a new queen.

2. Nucleus method: The nucleus method involves creating a small colony, or nucleus, from the existing colony. The nucleus is made up of several frames of brood, honey, and pollen, along with a queen cell or a new queen.

3. Walk-away split method: The walk-away split method involves dividing the hive into two unequal parts, with the majority of the bees and resources going to one part. The other part is left to raise a new queen and build up its own resources.

Tips for Successful Hive Splitting

1. Ensure that the colony is healthy and has enough resources to support the new colonies.

2. Make sure that each new colony has a queen bee and enough resources to support its growth.

3. Use high-quality equipment and tools for splitting the hive.

4. Follow the proper timeline for splitting the hive, and be careful not to disrupt the brood cycle.

5. Monitor the health and progress of each new colony, and make adjustments as needed.

Splitting hives and creating new colonies is an important part of beekeeping. By splitting healthy colonies, beekeepers can prevent swarming, create new colonies, manage disease, and maintain the health of the hive. There are several methods of splitting hives, each with its own advantages and disadvantages. With the right tools and techniques, hive splitting can be a successful and rewarding practice for beekeepers.

Beekeeping for pollination services

Beekeeping for pollination services is a vital part of modern agriculture. Many crops, such as almonds, apples, and blueberries, depend on honey bees for pollination. As a result, beekeepers can offer their services to farmers and earn income by providing pollination services.

In this section, we will discuss the basics of beekeeping for pollination services, including how it works, the benefits, and the challenges.

How Does Beekeeping for Pollination Services Work?

When a farmer hires a beekeeper for pollination services, the beekeeper will typically bring their hives to the farmer's field or orchard during the bloom period. The bees will then pollinate the crops as they forage for nectar and pollen.

The beekeeper will typically charge a fee for their services, based on the number of hives and the duration of the service. The fee can vary depending on the crop and the location.

Benefits of Beekeeping for Pollination Services

1. Increased Crop Yields

Bees are essential for pollinating many crops, including fruits, nuts, and vegetables. Many crops wouldn't bear fruit or would yield much less if bees weren't present. By providing pollination services, beekeepers can help increase crop yields and improve the quality of the harvest. This, in turn, benefits farmers by providing them with higher yields and increased profitability.

2. Improved Crop Quality

In addition to increasing crop yields, bees also improve the quality of the crops. Pollen is transferred from the male to the female portions of the flower by bees as they visit flowers in search of nectar and pollen. This process of pollination is essential for the development of healthy, robust fruits and vegetables. Pollinated crops are often larger, more flavorful, and have a longer shelf life.

3. Boosts Biodiversity

Beekeeping for pollination services also helps to promote biodiversity. By providing bees with a variety of pollen and nectar sources, beekeepers can help promote the growth of a diverse range of plants. This, in turn, supports a range of other insects and animals that rely on these plants for their own survival. A healthy and diverse ecosystem benefits everyone, including humans, by providing essential ecological services such as air and water purification, soil health, and pest control.

4. Provides Income for Beekeepers

Beekeeping for pollination services can be a profitable venture for beekeepers. They can earn income by providing their hives for pollination services during the bloom period. The fee for pollination services can vary depending on the crop and the location. For beekeepers, this can be an important source of income and can help to support their beekeeping operation.

5. Promotes Bee Health

Providing pollination services can be beneficial for the health of bee colonies. When bees are provided with a diverse range of pollen and nectar sources, they are less likely to suffer from nutritional deficiencies and stress. Additionally, by providing hives in different locations, beekeepers can help to reduce the risk of overcrowding and disease. Healthy bees are more productive, and this benefits beekeepers, farmers, and the ecosystem as a whole.

6. Supports Food Security

Beekeeping for pollination services is essential for food security. As the world's population grows, so does the demand for food. Bees play a vital role in the production of many crops, and without their pollination services, many crops would fail to produce fruit or would produce low yields. This, in turn, could lead to food shortages and increased food prices. By supporting beekeeping for pollination services, we are ensuring that we can meet the demands of a growing population and support food security.

Challenges of Beekeeping for Pollination Services

1. Transporting Hives

Transporting hives to different locations can be challenging and time-consuming. Beekeepers must ensure that the hives are securely loaded and transported in a way that minimizes stress on the bees. Additionally, beekeepers must ensure that they have the necessary permits and licenses to transport their hives, as regulations can vary from state to state.

2. Pesticide Exposure

Bees can be exposed to pesticides and other chemicals when they are used for pollination services. This exposure can be harmful to the bees and can have long-term effects on their health and productivity. It's important for beekeepers to work with farmers who use bee-friendly pest management practices and to monitor the health of the bees for signs of pesticide exposure.

3. Weather and Environmental Factors

Weather and environmental factors can affect the success of pollination services. Extreme temperatures, rainfall, and wind can all impact bee behavior and the effectiveness of pollination. For example, high winds can make it difficult for bees to fly and collect pollen, while heavy rainfall can wash away pollen from flowers. Beekeepers must be prepared to adapt to these conditions and make adjustments as needed.

4. Colony Health

Beekeeping for pollination services can put additional stress on bee colonies. Bees are often transported long distances and exposed to new environments and pollen sources. This can lead to increased stress and disease in the colonies. Beekeepers must monitor the health of the colonies and take preventative measures and treat disease.

5. Hive Theft

Hive theft is a growing concern for beekeepers. Hives can be expensive, and their value as pollination services providers can make them a target for theft. Beekeepers must take steps to secure their hives and prevent theft, such as using GPS tracking devices and locking their hives.

6. Labor Costs

Beekeeping for pollination services requires a significant amount of labor. Beekeepers must transport their hives to different locations, monitor the health of the colonies, and manage the overall operation.

These labor costs can be significant and can impact the profitability of the operation.

Tips for Successful Beekeeping for Pollination Services

1. Work with Bee-Friendly Farmers

It's important to work with farmers who use bee-friendly pest management practices and who are committed to protecting the environment. Bee-friendly farmers use natural and sustainable pest management practices that minimize the risk of pesticide exposure to bees. Additionally, they provide a diverse range of pollen and nectar sources that help to promote bee health and productivity.

2. Monitor the Health of the Bees

It's important to monitor the health of the bees for signs of stress or disease. Regular hive inspections and monitoring of brood and honey stores can help ensure the health of the colony. Beekeepers should be familiar with the signs of common diseases and pests that affect bees, such as varroa mites and American foulbrood, and take proactive steps to prevent and treat them.

3. Choose the Right Hives

The number and size of hives used for pollination services will depend on the crop and the location. Beekeepers should choose the right hives for the job, and ensure that they are in good condition and well-stocked with bees and resources. Additionally, beekeepers should ensure that the hives are securely loaded and transported in a way that minimizes stress on the bees.

4. Provide Adequate Nutrition

Providing bees with a diverse range of pollen and nectar sources is essential for their health and productivity. Beekeepers should ensure that their hives have access to a variety of flowering plants that bloom throughout the season. Additionally, beekeepers can provide supplemental nutrition, such as sugar syrup or pollen patties, during times of low nectar flow.

5. Manage Colony Size

Beekeepers should manage the size of their colonies to ensure that they are strong and healthy. Overcrowding can lead to increased stress and disease in the colony. Beekeepers can prevent overcrowding by splitting hives or adding supers to accommodate the growing population.

6. Be Prepared for Weather and Environmental Factors

Weather and environmental factors can impact the success of pollination services. Beekeepers should be prepared to adapt to these conditions and make adjustments as needed. For example, providing shade during hot weather or windbreaks during windy weather can help to protect the hives and ensure the health of the bees.

7. Work with a Mentor

Beekeeping for pollination services can be a complex and challenging venture. Working with a mentor or experienced beekeeper can help beekeepers learn the skills and techniques needed for success. Mentors can provide guidance on hive management, disease prevention, and other aspects of beekeeping.

How to make beeswax candles and soap?

Beeswax candles and soap are popular natural products that have been used for centuries. Beeswax is a natural wax produced by honey bees and is known for its distinctive fragrance, smooth texture, and ability to burn cleanly. Making beeswax candles and soap is a simple and rewarding process that can be done at home with the right materials and techniques.

How to Make Beeswax Candles

Materials Needed:

- Beeswax

- Candle wick

- Wick clips or adhesive

- Double boiler or melting pot

- Thermometer

- Candle molds (optional)

- Essential oils (optional)

Instructions:

1. Melt the beeswax: Use a double boiler or melting pot to melt the beeswax. Beeswax melts at a temperature of approximately 145-150 degrees Fahrenheit. Keep an eye on the wax's temperature with a thermometer.

2. Prepare the wick: Trim the wick to the appropriate length and attach it to the bottom of the candle mold or container using

wick clips or adhesive. Make sure the wick is centered in the mold or container.

3. Pour the wax: Once the beeswax has melted, remove it from the heat and allow it to cool slightly. Add any essential oils if desired. Slowly pour the wax into the candle mold or container, making sure the wick stays centered.

4. Allow to cool: Allow the wax to cool completely and harden. Depending on the size of the candle, this may require many hours.

5. Trim the wick: Trim the wick to the appropriate length when the wax has cooled and hardened.

Tips:

- Beeswax can be colored using natural dyes such as beetroot powder or turmeric.

- Beeswax candles burn cleaner and longer than paraffin wax candles.

- Beeswax candles have a natural fragrance that can be enhanced with the addition of essential oils.

Types of Beeswax Candles

There are several different types of beeswax candles that can be made using beeswax as the main material. Here are some of the most popular types:

1. Rolled Beeswax Candles: These candles are made by rolling a sheet of beeswax around a wick. The sheets of beeswax are often sold in pre-cut sizes, and the size of the finished candle depends on the size of the sheet used. Rolled beeswax candles are simple to make and can be decorated with different colors and patterns.

2. Pillar Beeswax Candles: Pillar candles are made by pouring melted beeswax into a mold and allowing it to cool and harden. These candles can be made in a variety of shapes and sizes, including round, square, and cylindrical. They can also be decorated with different colors and patterns by adding layers of different colored beeswax.

3. Votive Beeswax Candles: Votive candles are small candles that are usually made in small containers or molds. They are typically burned inside larger candle holders or lanterns. Beeswax votive candles are made by pouring melted beeswax into small containers and allowing it to cool and harden.

4. Tealight Beeswax Candles: Tealight candles are small, circular candles that are commonly used in decorative candle holders. Beeswax tealight candles are made by pouring melted beeswax into a circular mold and allowing it to cool and harden. They can also be decorated with different colors and patterns.

5. Taper Beeswax Candles: Long, thin candles known as "taper candles" are frequently used in candlesticks or candelabras.

Beeswax taper candles are made by pouring melted beeswax into a taper mold and allowing it to cool and harden. The molds are typically tapered to create the desired shape.

How to Make Beeswax Soap

Materials Needed:

- Beeswax

- Coconut oil

- Olive oil

- Lye

- Distilled water

- Essential oils (optional)

- Soap molds

Instructions:

1. Melt the beeswax and oils: Melt the beeswax, coconut oil, and olive oil in a double boiler or melting pot. Beeswax melts at a temperature of approximately 145-150 degrees Fahrenheit.

2. Prepare the lye solution: Mix the lye and distilled water in a separate container, stirring until dissolved. Be sure to use caution and wear protective gloves and eyewear when working with lye, as it is a strong and potentially dangerous chemical.

3. Combine the two mixtures: Slowly pour the lye solution into the melted beeswax and oils, stirring continuously until well combined. Add any essential oils if desired.

4. Pour into molds: Pour the soap mixture into soap molds and allow to cool and harden. The molds can be created using silicone, plastic, or metal.

5. Remove from molds: Once the soap has cooled and hardened, remove it from the molds and cut into bars. Allow the soap to cure for several weeks before using it, as this will allow the soap to harden and become more effective.

Tips:

- Beeswax soap has a creamy texture and provides a rich lather.

- Beeswax soap can be scented with essential oils for added fragrance.

- Beeswax soap is gentle on the skin and is suitable for people with sensitive skin.

By following these instructions, you can create beautiful and natural beeswax candles and soap for personal use or to give as gifts.

CHAPTER
VI
Troubleshooting
and Problem Solving

Common challenges faced by beekeepers

Beekeeping is a rewarding and fulfilling hobby or profession, but it is not without its challenges. Here are some of the common challenges faced by beekeepers.

1. Colony Health

One of the most significant challenges faced by beekeepers is managing the health of their colonies. Bees are susceptible to a range of diseases and pests, including varroa mites, American foulbrood, and chalkbrood. Additionally, colony collapse disorder (CCD) is a serious concern for beekeepers, as it can lead to the loss of entire colonies. Beekeepers must monitor their colonies regularly for signs of disease or pest infestations and take proactive steps to prevent and treat them.

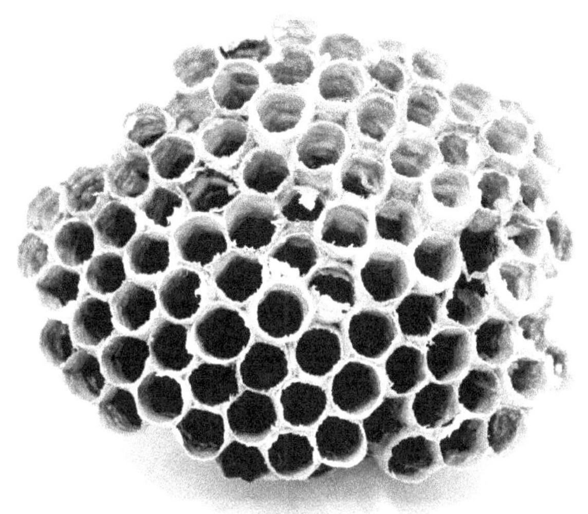

2. Environmental Factors

Environmental factors can have a significant impact on bee behavior and colony health. Factors such as extreme temperatures, drought, and rainfall can impact the availability of food sources and the bees' ability to forage. Additionally, exposure to pesticides and other chemicals can be harmful to bees. Beekeepers must be aware of the environmental factors that impact their colonies and take steps to mitigate their effects.

3. Queen Health and Reproduction

The queen bee is a vital part of the colony's reproductive system. If the queen bee is unhealthy or fails to produce enough eggs, the colony can suffer. Beekeepers must ensure that their queen bees are healthy and productive. This involves regular inspections of the queen and monitoring the brood pattern to ensure that the colony is producing enough young bees.

4. Honey Production

Honey production is an essential aspect of beekeeping for many beekeepers. However, producing high-quality honey requires careful management of the colonies. Beekeepers must ensure that their colonies have enough food to sustain themselves, while also leaving enough honey for the beekeeper to harvest. Additionally, beekeepers must harvest honey at the right time to ensure that it is of the highest quality.

5. Bee Behavior

Understanding bee behavior is essential for successful beekeeping. Bees can be unpredictable and can exhibit unexpected behavior. Beekeepers must be able to read the behavior of their colonies and respond accordingly. For example, aggressive behavior may indicate that the colony is stressed or diseased, while swarming behavior may indicate that the colony is overcrowded.

6. Regulation and Legislation

Beekeeping is subject to a range of regulations and legislation. Beekeepers must ensure that they are compliant with local, state, and federal laws. This may involve obtaining permits, registering their colonies, and following specific guidelines for beekeeping operations.

7. Financial Considerations

Beekeeping can be an expensive venture. Beekeepers must purchase hives, protective gear, and other equipment. Additionally, managing the health of the colonies can be costly, as it may involve purchasing medications and supplements. Beekeepers must carefully manage their finances to ensure the success of their operation.

Beekeeping is a challenging but rewarding endeavor. Beekeepers must navigate a range of issues, from managing the health of their colonies to dealing with environmental factors that impact bee behavior. By staying informed about the common challenges faced by beekeepers and taking proactive steps to address them, beekeepers can ensure the success of their operation and the health of their colonies. It's essential for beekeepers to continually educate themselves about beekeeping best practices and to seek guidance from experienced beekeepers and mentors.

Identifying and addressing problems in your hive

Identifying and addressing problems in your hive is an essential aspect of beekeeping. Bees are susceptible to a range of diseases and pests, and it's important for beekeepers to monitor their colonies regularly to ensure the health and productivity of their bees. In this section, we will discuss in detail how to identify and address problems in your hive.

1. Conduct Regular Inspections

Conducting regular inspections of your hive is the first step in identifying and addressing problems. Inspections should be done every 1-2 weeks during the peak season and less frequently during the winter months. During inspections, beekeepers should look for signs of disease or pest infestations, as well as ensure that the hive has enough food and space.

2. Recognize the Signs of Disease

Beekeepers should be familiar with the signs of common diseases that affect bees, such as varroa mites, American foulbrood, and chalkbrood. Signs of disease can include abnormal brood patterns, deformed wings, or the presence of dead or dying bees. If you suspect that your hive is infected with a disease, it's important to take immediate action to prevent the spread of the disease to other colonies.

3. Manage Pests

Pests such as varroa mites, wax moths, and small hive beetles can have a significant impact on the health and productivity of a hive. Beekeepers should monitor their colonies regularly for signs of pest infestations, such as small holes in the comb or the presence of larvae. If pests are detected, beekeepers should take immediate action to control the infestation and prevent it from spreading to other colonies.

4. Maintain Adequate Nutrition

Maintaining adequate nutrition is essential for the health and productivity of a hive. Bees require a diverse range of pollen and nectar sources to maintain their health and produce honey. If a hive is not receiving enough food, beekeepers should consider providing supplemental nutrition, such as sugar syrup or pollen patties.

5. Manage Colony Size

Overcrowding can lead to increased stress and disease in a hive. Beekeepers should monitor the size of their colonies and take steps to prevent overcrowding, such as splitting the hive or adding supers to accommodate the growing population. Additionally, beekeepers should ensure that their hives have enough space to store honey and brood.

6. Address Environmental Factors

Environmental factors can have a significant impact on the health and behavior of bees. Extreme temperatures, drought, and rainfall can all impact the availability of food sources and the bees' ability to forage. Additionally, exposure to pesticides and other chemicals can be harmful to bees. Beekeepers should be aware of the environmental factors that impact their colonies and take steps to mitigate their effects.

7. Seek Professional Assistance

If you are unsure how to identify or address a problem in your hive, it's important to seek professional assistance. Local beekeeping associations or extension offices can provide guidance and support to beekeepers. Additionally, experienced beekeepers or mentors can provide valuable advice and assistance.

Beekeeping in adverse weather conditions

Adverse weather conditions can take many forms, such as extreme heat or cold, high winds, heavy rain, and snow. These conditions can have a significant impact on the health and productivity of bee colonies. For example, extreme heat can cause dehydration, leading to a lack of nectar and pollen resources, while extreme cold can freeze bees and disrupt their brood rearing activities.

Cold Weather

One of the most common adverse weather conditions beekeepers face is cold weather. Cold weather can be detrimental to bees because it can cause them to cluster together in the hive to stay warm, which can deplete their food resources quickly. Bees need to consume honey to generate heat, and if they run out of honey, they can starve to death. Therefore, it is crucial for beekeepers to ensure that their hives have adequate food reserves before the onset of cold weather.

Beekeepers can also use insulation to help keep their hives warm during cold weather. Insulation can help prevent the depletion of honey reserves and keep bees healthy and productive. Beekeepers can use materials such as foam board or straw to insulate their hives. However, it is important to ensure that the insulation does not block the entrance to the hive, as this can cause ventilation issues.

Humidity

Another challenge that beekeepers face in adverse weather conditions is maintaining the proper humidity levels in the hive. In humid conditions, bees can become susceptible to fungal and bacterial infections, which can be deadly for the colony. On the other hand, in dry conditions, bees can become dehydrated and unable to produce enough honey to sustain themselves.

Beekeepers should regularly check the humidity levels in their hives and take measures to maintain a healthy balance. One way to do this is by using a hygrometer to measure the humidity levels in the hive. A healthy humidity level for a bee hive is between 40% and 60%. If humidity levels are too high, beekeepers can use a dehumidifier or increase ventilation to reduce humidity. If humidity levels are too low, beekeepers can provide supplemental water sources such as moistened sponges or small bowls of water in the hive.

High Winds

High winds are another adverse weather condition that beekeepers need to be aware of. Strong winds can topple hives and disrupt the flight patterns of bees, making it difficult for them to forage for food. Beekeepers can mitigate the effects of high winds by ensuring that their hives are securely anchored to the ground and providing windbreaks, such as trees or fences.

Beekeepers can also use hive straps or weights to secure their hives during high winds. This can prevent the hive from toppling over and causing damage to the colony. Additionally, beekeepers can provide extra ventilation during high winds to prevent the buildup of moisture inside the hive.

Heavy Rain and Flooding

Heavy rain and flooding can also pose significant challenges for beekeepers. Prolonged periods of rain can wash away nectar and pollen resources, leaving bees with limited food options. Floods can also drown bees and damage hives, leading to colony loss.

Beekeepers can prepare for heavy rain and flooding by ensuring that their hives are located in areas with good drainage and elevating hives off the ground. Beekeepers can also provide extra food reserves to their colonies during periods of heavy rain to ensure that they have enough food to survive.

Extreme Weather Events

In addition to these challenges, beekeepers may also face other adverse weather conditions, such as droughts, wildfires, and hurricanes. These conditions can be unpredictable and often require advanced preparation and planning to mitigate their effects.

Beekeepers should have a plan in place for evacuating their hives in the event of an extreme weather event. This may involve relocating hives to a safe location or covering hives with protective materials such as tarps or plywood. It is also important to monitor weather conditions closely and take appropriate measures to protect hives and bees from the effects of extreme weather events.

To manage beekeeping in adverse weather conditions, beekeepers need to be proactive and take steps to ensure that their hives are healthy and productive. Here are some tips on how to manage beekeeping in adverse weather conditions:

1. Inspect your hives regularly: Regular inspections are crucial for identifying potential issues before they become major problems. During adverse weather conditions, it is even more important to inspect your hives frequently for any signs of damage or distress.

2. Provide adequate food: Bees need sufficient food reserves to survive during periods of low food availability, such as during droughts or heavy rain. Ensure that your hives have enough honey and pollen reserves to sustain the colony. You can also provide supplemental feeding during these times, such as sugar syrup or pollen patties.

3. Manage humidity levels: Bee colonies need a balanced humidity level to thrive. High humidity levels can lead to the growth of bacteria and fungi, while low humidity levels can cause dehydration. Use a hygrometer to measure the humidity levels in your hives and take measures to maintain a healthy balance.

4. Protect your hives from high winds: High winds can cause hives to topple over and disrupt the flight patterns of bees. Securely anchor your hives to the ground and provide windbreaks, such as trees or fences, to protect your bees from the effects of high winds.

5. Protect your hives from flooding: During heavy rain or flooding, hives can become waterlogged, leading to colony loss. Elevate your hives off the ground or locate them in areas with good drainage to prevent water damage.

6. Keep your hives warm during cold weather: Bees need warmth to survive during cold weather. Use insulation to keep your hives warm and prevent the depletion of honey reserves. Insulate the walls and roof of the hive, but ensure that the entrance remains unobstructed for ventilation.

7. Be prepared for extreme weather events: Prepare for extreme weather events, such as hurricanes or wildfires, by having a plan in place for evacuating your hives if necessary. Take steps to protect your hives from damage, such as covering them with tarps or plywood.

8. Practice good hygiene: During adverse weather conditions, diseases and pests can spread more easily. Practice good

hygiene by keeping your hives clean and free of debris. This can help prevent the spread of disease and maintain healthy colonies.

9. Monitor your bees' behavior: Bees are sensitive to changes in their environment and will display signs of stress during adverse weather conditions. Monitor your bees' behavior and take measures to alleviate stress, such as providing supplemental feeding or reducing hive traffic.

10. Stay informed: Stay informed about local weather conditions and potential weather hazards. This will help you prepare for adverse weather conditions and take appropriate measures to protect your bees. You can also join local beekeeping associations or online forums to stay up-to-date on best practices and tips for managing beekeeping in adverse weather conditions.

In conclusion, beekeeping in adverse weather conditions can be challenging, but with proper preparation and management, beekeepers can maintain healthy and productive hives. Beekeepers should be proactive in identifying potential issues and taking steps to prevent or mitigate their effects. By following the tips outlined in this section, beekeepers can ensure the survival of their hives and enjoy the many benefits of beekeeping, even in adverse weather conditions.

CHAPTER
VII
Marketing and Selling Honey and Other Bee Products

Selling honey and other bee products

Selling honey and other bee products is a great way for beekeepers to earn additional income and share the many benefits of beekeeping with others. Bee products such as honey, beeswax, pollen, and propolis have a variety of uses, from cooking and baking to natural skincare and medicinal remedies. In this section, we will explore the various aspects of selling honey and other bee products, including marketing, packaging, pricing and other strategies.

1. Marketing: Marketing bee products involves promoting the unique qualities and benefits of your products to potential customers. One of the most effective ways to market bee products is through word-of-mouth advertising. Encourage your satisfied customers to share their positive experiences with others, whether through social media, online reviews, or in-person recommendations.

   Another effective marketing strategy is to participate in local farmers' markets or craft fairs. These events provide a great opportunity to showcase your bee products to a wider audience and connect with potential customers. You can also consider creating an online store or selling through third-party platforms such as Etsy or Amazon.

2. Packaging: Packaging bee products is an important aspect of selling bee products. The packaging should be both functional and visually appealing to attract customers. Honey should be packaged in a clean, sealed container, such as a glass jar, with a label that includes the type of honey, the weight, and any other relevant information, such as the location of the hive or the harvesting date.

   For beeswax products, such as candles or balms, consider using environmentally friendly packaging materials, such as paper or cardboard. You can also include information on the benefits of beeswax and how it is produced by bees.

3. Pricing: Pricing bee products can be a challenging task, as there are many factors to consider, including the cost of production, market demand, and competition. It is important to ensure that your prices are competitive and reflect the value of your products.

   One way to determine the pricing of your bee products is to research the prices of similar products in your area. Consider the quality and size of the products, as well as any unique features or benefits that your products may offer.

   You can also factor in the cost of production, such as the cost of equipment, labor, and supplies. It is important to ensure that your prices cover the cost of production, as well as provide a reasonable profit margin.

4. Sales channels: There are various sales channels available for selling bee products, including farmers' markets, craft fairs, online stores, and third-party platforms such as Etsy or Amazon. Each sales channel has its own advantages and disadvantages, and it is important to choose the right sales channel that fits your business needs.

   For example, farmers' markets and craft fairs provide an opportunity to connect with customers face-to-face and build relationships, while online stores and third-party platforms allow you to reach a wider audience and provide convenience for customers.

5. Legal requirements: Selling bee products may be subject to local, state, and federal regulations, including food safety regulations, labeling requirements, and tax laws. It is important to research and comply with all applicable laws and regulations to avoid any legal issues.

For example, honey may be subject to food safety regulations, and may require testing for certain contaminants such as antibiotics and pesticides. You may also need to obtain a food handler's permit or license to sell honey and other bee products.

Types of Bee Products to Sell

There are numerous varieties of bee products that beekeepers can sell, including:

1. Honey: Honey is the most well-known and widely consumed bee product. It is a sweet, viscous substance produced by honeybees

from the nectar of flowers. The taste, color, and texture of honey can vary depending on the type of flower that the bees visited, as well as the climate and soil conditions. Some popular types of honey include clover honey, wildflower honey, and honey from specific plants or regions. Honey is used for a variety of purposes, including cooking, baking, and sweetening beverages.

2. Beeswax: A natural wax made by honeybees is called beeswax. It is used by bees to build their comb and store honey and pollen. Beeswax is a versatile product that can be used for a variety of purposes, including candles, lip balms, and skincare products. It is also used as an ingredient in furniture polish and shoe polish.

3. Pollen: Bee pollen is a nutrient-rich substance collected by bees from the stamens of flowers. It contains vitamins, minerals, and protein, and is often sold in granule or powder form. Bee pollen is thought to have a variety of health advantages, such as lowering inflammation, enhancing the immune system, and increasing energy. It can be added to smoothies, yogurt, or other foods.

4. Propolis: Propolis is a sticky resin produced by honeybees from the sap of trees and plants. Bees use propolis to seal cracks and gaps in the hive, and to protect against bacteria and other pathogens. Propolis has antibacterial and antifungal properties and is often used in natural remedies and skincare products. It is sold in various forms, including tinctures, capsules, and creams.

5. Royal jelly: Royal jelly is a nutrient-rich substance produced by worker bees and fed to queen bees. It is a milky white substance with a slightly sour taste. Royal jelly contains vitamins, minerals, and amino acids, it is thought to provide a variety of health advantages, including strengthening the immune system and reducing inflammation. It is often sold in capsule or liquid form as a dietary supplement.

6. Mead: Mead is an alcoholic beverage made from fermented honey. It has been enjoyed for thousands of years and is often considered to be the world's oldest alcoholic beverage. Mead can be sweet or dry, and can be flavored with fruits, spices, or herbs. It is a popular alternative to beer or wine and can be sold in bottles or kegs.

In conclusion, selling honey and other bee products can be a rewarding and profitable endeavor for beekeepers.

Building a brand and finding your target market

Building a brand and finding your target market is an important aspect of selling honey and other bee products. A strong brand can help differentiate your products from competitors and build customer loyalty, while targeting the right market can ensure that your products are reaching the customers who are most likely to purchase them. In this section, we will explore the various aspects of building a brand and finding your target market when selling honey and other bee products.

Building a Brand

Building a brand involves creating a unique identity for your business and products. A strong brand can help your products stand out in a crowded market, build customer loyalty, and increase sales. Here are some key steps to building a strong brand for your bee products:

1. Define Your Brand Values: The first step to building a strong brand for your bee products is to define your brand values. This involves identifying the core values and mission of your business, and how these values are reflected in your products. What sets your bee products apart from others? What are the key benefits of your products? Understanding your brand values will help guide your marketing and sales strategies.

2. Develop a Brand Identity: Once you have defined your brand values, the next step is to develop a visual identity for your brand. This includes creating a logo, color scheme, typography, and other visual elements that will be used across all marketing materials and product packaging. Your brand identity should reflect the values of your business and appeal to your target market.

3. Create a Brand Story: Creating a brand story is an important aspect of building a strong brand for your bee products. Your brand story should connect your brand values to the products you sell. Tell the story of your bees and how they produce your products. Highlight the natural and sustainable aspects of your products, as well as any unique features or benefits. Creating a

compelling brand story can help differentiate your products from competitors and build customer loyalty.

4. Build Relationships: Building relationships with customers is an important aspect of building a strong brand for your bee products. Connect with customers through social media, email newsletters, and other channels. Respond to customer inquiries and feedback in a timely and friendly manner. Building relationships with customers can help build trust and increase customer loyalty.

5. Offer Quality Products and Services: Offering quality products and services is essential to building a strong brand for your bee products. Ensure that your bee products are of high quality and meet customer expectations. Provide excellent customer service and ensure that orders are delivered in a timely manner. Offering quality products and services can help build customer loyalty and generate positive word-of-mouth advertising.

6. Engage in Sustainable and Responsible Practices: Engaging in sustainable and responsible practices is another important aspect of building a strong brand for your bee products. Ensure that your beekeeping practices are sustainable and do not harm the environment. Use environmentally friendly packaging materials and minimize waste. Engaging in sustainable and responsible practices can help differentiate your products from competitors and appeal to environmentally conscious customers.

7. Partner with Other Businesses and Organizations: Partnering with other businesses and organizations can help build a strong brand for your bee products. Look for opportunities to collaborate with other businesses and organizations that share your values and mission. For example, you could partner with local farmers' markets or craft fairs, or donate a portion of your profits to environmental organizations. Partnering with other businesses and organizations can help expand your reach and build your brand reputation.

Finding Your Target Market

Finding your target market involves identifying the customers who are most likely to purchase your products. By understanding your target market, you can tailor your marketing and sales strategies to reach the right customers. Here are some key steps to finding your target market for bee products:

1. Define Your Ideal Customer: The first step to finding your target market for bee products is to define your ideal customer. This involves identifying the demographic and psychographic characteristics of the customers who are most likely to purchase your products. Consider factors such as age, gender, income, geographic location, values, interests, and purchasing behaviors.

   For example, if you produce organic honey, your ideal customer may be health-conscious consumers who are interested in natural and organic products. If you produce beeswax candles, your ideal customer may be environmentally conscious consumers who value sustainability and natural materials.

2. Research the Market: The next step to finding your target market for bee products is to research the market for bee products in your area. This includes understanding the demand for honey, beeswax, and other bee products, as well as identifying any gaps in the market that your products can fill. Look for trends and patterns in the market, and consider how your products can meet the needs of customers.

   For example, if there is a high demand for local, organic honey in your area, you may want to focus your marketing efforts on promoting the natural and sustainable aspects of your honey. Alternatively, if there is a gap in the market for beeswax candles, you may want to expand your product line to include a variety of scents and designs that appeal to different customer preferences.

3. Analyze Customer Data: Analyzing customer data can provide valuable insights into your target market. Look at customer purchase history, feedback, and social media activity to understand their preferences and behaviors. Use this data to identify patterns and trends in customer behavior, and tailor your marketing and sales strategies accordingly.

   For example, if you notice that customers are more likely to purchase honey during certain times of the year, such as the holiday season, you may want to increase your marketing efforts during those times. Alternatively, if customers are more likely to purchase beeswax candles in specific scents or designs, you may want to expand your product line to include those options.

4. Tailor Marketing and Sales Strategies: Once you have identified your target market for bee products, the final step is to tailor your marketing and sales strategies to reach that market. Use marketing channels and messaging that resonate with your target market, and offer promotions and discounts that appeal to their interests and preferences.

For example, if your target market is health-conscious consumers, you may want to focus your marketing efforts on promoting the natural and organic aspects of your honey, and offer discounts or promotions for repeat customers. Alternatively, if your target market is environmentally conscious consumers, you may want to focus your marketing efforts on the sustainable and eco-friendly aspects of your products, and partner with other environmentally focused businesses and organizations.

Honey packaging and labeling regulations

Honey is a natural product that is widely consumed across the world due to its numerous health benefits. It is a sweet, viscous substance produced by bees from the nectar of flowers. Honey is used as a food ingredient and as a natural sweetener in many foods and beverages. Honey is also used in medicine and skincare products. Honey packaging and labeling regulations are important because they ensure that consumers are informed about the product they are buying and that the honey is safe for consumption.

Honey packaging and labeling regulations are guidelines and requirements put in place by regulatory bodies to ensure the safety, quality, and integrity of honey products. These regulations specify the materials, design, labeling, and other requirements that honey producers and packagers must adhere to in order to maintain compliance with the law and ensure the safety of consumers.

Packaging Regulations for Honey

Honey is a delicate and perishable product that requires careful handling and packaging to preserve its freshness, quality, and safety. Honey packaging regulations outline the requirements for the packaging materials, design, and labeling that honey producers and packagers must follow to ensure that their products meet regulatory standards.

Packaging Materials

Honey packaging materials must be made of food-grade materials that do not react with or contaminate the honey. Glass jars are the most common packaging material for honey as they are non-reactive, non-porous, and do not affect the flavor or quality of the honey. Plastic containers may also be used for honey packaging, but they must be made of food-grade plastic that does not contain any harmful chemicals.

Packaging Design

Honey packaging must be designed to protect the honey from contamination, moisture, and other external factors that can affect its quality and safety. Honey containers must be sealed tightly to prevent leakage, and the packaging design must ensure that the honey is easily accessible to consumers.

Labeling Regulations for Honey

Honey labeling regulations specify the information that must be included on the label of honey products to ensure that consumers are informed about the product they are buying. Honey labeling regulations may vary by country, but they typically include the following information:

Product Name: The product name must clearly state that the product is honey. The use of any other name or term to describe the product is prohibited, except for the addition of a floral or geographic designation.

Origin: The label must state the country of origin of the honey. If the honey is a blend of honey from different countries, the label must state the percentage of honey from each country.

Net Quantity: The label must state the net quantity of honey in the package in metric units. The net quantity must be accurate and must not include the weight of the packaging material.

Date of Minimum Durability: The label must state the date of minimum durability, also known as the "best before" date. This is the date after which the honey may start to lose its quality or spoil. The date of minimum durability must be indicated using the day, month, and year.

Instructions for Use: The label must provide instructions for the use of the honey. This includes information on how to store the honey and how to use it.

Allergen Information: The label must state if the honey contains any allergens such as nuts, soy, or gluten.

Nutritional Information: The label must state the nutritional information of the honey. This includes the amount of calories, carbohydrates, and sugars in the honey.

Health Claims: The label must not make any false or misleading health claims about the honey. Health claims must be supported by scientific evidence.

Organic Labeling: Honey that is labeled as "organic" must be certified by a recognized organic certifying body. The label must state the name of the certifying body and the organic certification number.

In conclusion, Honey packaging and labeling regulations are essential for ensuring the safety, quality, and integrity of honey products. These regulations provide guidance on the packaging materials, design, and labeling requirements that honey producers and packagers must follow to ensure that their products meet regulatory standards. By following honey packaging and labeling regulations, honey producers can ensure that their products are of high quality and safe for consumption. Consumers can also have confidence in the safety and quality of the honey they are purchasing, knowing that it meets regulatory standards.

Creating a successful business plan for beekeeping

Beekeeping is an exciting and rewarding business opportunity that has gained popularity in recent years. A successful beekeeping business requires a solid business plan that outlines the key strategies, goals, and objectives of the business. A well-written business plan provides a roadmap for success and helps to secure funding from investors or financial institutions. In this section, we will discuss the essential elements of a successful business plan for beekeeping.

Executive Summary

The executive summary is one of the essential elements of a successful business plan for beekeeping. It is the first section of the business plan and provides a brief overview of the entire plan. The purpose of the executive summary is to provide a snapshot of the business plan and to give the reader an understanding of the key components of the plan. It should be written in a way that is easily understandable by anyone who reads it, including potential investors or financial institutions.

The executive summary should include the following information:

1. Business Overview: A brief description of the business, including the products or services offered, the target market, and the competitive advantage of the business.

2. Goals and Objectives: The goals and objectives of the business, including revenue projections, growth targets, and market share.

3. Management Team: A summary of the management team, including their experience and qualifications.

The executive summary should be no more than one or two pages in length.

Company Description

The company description is an important element of a successful business plan for beekeeping because it sets the tone for the entire plan. It provides a detailed overview of the business, including its mission, vision, and values. The purpose of the company description is to provide potential investors or financial institutions with a clear understanding of the business and what it stands for.

The company description should include the following information:

1. Legal Structure: The legal structure of the business, including whether it is a sole proprietorship, partnership, or corporation.

2. Location: The location of the business, including any physical facilities such as apiaries, warehouses, or offices.

3. Products or Services: A description of the products or services offered by the business, including the types of honey and other bee products sold, such as beeswax, pollen, and propolis.

4. Mission, Vision, and Values: A clear statement of the business's mission, vision, and values. This should include the purpose of the business, its long-term goals, and the values that guide its operations.

Market Analysis

Market analysis is one of the essential elements of a successful business plan for beekeeping because it helps to identify the market opportunities and challenges facing the business. It provides an overview of the industry and market in which the business operates.

The purpose of the market analysis is to provide a clear understanding of the market, including trends, challenges, and opportunities, as well as the target market and competition.

The market analysis should include the following information:

1. Industry Analysis: An overview of the beekeeping industry, including its size, growth trends, and key players. This should also include an analysis of the challenges and opportunities facing the industry.

2. Target Market: An analysis of the target market for the business, including demographics, psychographics, and purchasing behavior. This should also include an analysis of the needs and preferences of the target market.

3. Competition: An analysis of the competition, including their strengths, weaknesses, and market share. This should also include an analysis of the competitive landscape and the key players in the industry.

Marketing and Sales Strategy

Marketing and sales strategy is one of the essential elements of a successful business plan for beekeeping. It outlines the strategies and tactics that the business will use to reach its target market and achieve its revenue goals. The purpose of the marketing and sales strategy is to provide a clear understanding of how the business will market and sell its products or services.

The marketing and sales strategy should include the following information:

1. Marketing Channels: A description of the marketing channels that the business will use to reach its target market. This can include social media, online advertising, or local farmers' markets.

2. Pricing Strategy: A description of the pricing strategy, including the prices of the products and any discounts or promotions. This should also include an analysis of the pricing strategies of the competition.

3. Sales Process: A description of the sales process, including how the products will be sold and distributed. This should include information about the sales team, the distribution channels, and any third-party vendors.

Operations and Management Plan

The Operations and Management Plan is one of the essential elements of a successful business plan for beekeeping. It provides an overview of the day-to-day operations of the business and the management structure. The purpose of the Operations and Management Plan is to provide a clear understanding of how the business will be managed and how it will operate.

The Operations and Management Plan should include the following information:

1. Production Process: A description of the production process, including how the bees will be managed and how the honey and other bee products will be extracted and processed. This should also include information about the quality control measures that will be implemented.

2. Management Structure: A description of the management structure, including the roles and responsibilities of each member of the team. This should include information about the experience and qualifications of each member of the team.

3. Systems and Processes: A description of the systems and processes that the business will use to manage its operations, such as inventory management, quality control, and customer service.

Financial Plan

The Financial Plan is one of the essential elements of a successful business plan for beekeeping. It provides a detailed overview of the financial projections for the business, including the startup costs, revenue streams, cash flow projections, and funding requirements. The purpose of the Financial Plan is to provide a clear understanding of the financial viability of the business.

The Financial Plan should include the following information:

1. Startup Costs: An estimation of the startup costs, including the costs of equipment, supplies, and any necessary permits or licenses.

2. Revenue Streams: A description of the revenue streams for the business, including the sales of honey, beeswax, and other bee products, as well as any income from pollination services.

3. Cash Flow Projections: Projections of the cash flow for the business, including the expected income and expenses for the first few years of operation. This should include a detailed analysis of the revenue and expenses for each month or quarter.

4. Funding Requirements: A description of the funding requirements for the business, including any loans, grants, or investments needed to start or grow the business.

CONCLUSION

In conclusion, "Beekeeping for Beginners - A Comprehensive Beginner's Guide to Bee Farming" provides a wealth of information for anyone interested in starting their own beekeeping operation. From the basics of bee anatomy and behavior, to choosing the right hive, to harvesting honey and beeswax products, this guide covers all aspects of beekeeping in an accessible and easy-to-understand way.

By following the steps outlined in this guide, beginners can confidently start their own beekeeping operation, with the knowledge and tools needed to ensure the health and well-being of their bees. Additionally, the guide covers important topics such as recognizing and managing common problems and diseases, feeding your bees, and how to split a hive or catch a swarm.

Whether you're interested in beekeeping as a hobby or as a business venture, "Beekeeping for Beginners" provides the necessary guidance to help you get started. By understanding the intricacies of beekeeping, you'll gain a new appreciation for the importance of bees in our ecosystem, and the many benefits that can be gained from beekeeping, including the production of honey, beeswax, and other natural products.

In addition to the practical knowledge provided in this guide, readers will also gain a deeper understanding of the history and cultural significance of beekeeping. The guide highlights the importance of bees in human history, from ancient civilizations to modern times, and the role that beekeeping has played in shaping our understanding of the natural world.

Overall, "Beekeeping for Beginners - A Comprehensive Beginner's Guide to Bee Farming" is an essential resource for anyone interested in beekeeping. Whether you're a seasoned beekeeper looking to expand your knowledge, or a beginner just starting out, this guide provides the knowledge and tools needed to create a successful and thriving beekeeping operation. With its accessible writing style and comprehensive coverage of all aspects of beekeeping, this guide is a must-read for anyone interested in this fascinating and important field.

Tips for a successful experience in beekeeping

If you're interested in starting a beekeeping operation, here are some tips for a successful experience:

1. Educate Yourself: Before starting a beekeeping operation, it is important to educate yourself on the basics of bee biology, hive management, and honey production. There are many resources available, including books, online courses, and local beekeeping associations. By understanding the needs and behaviors of bees, you can better manage your hives and ensure the health and productivity of your bees.

2. Start Small: It's important to start small, especially if you're new to beekeeping. Begin with a single hive, and learn as you go. As you gain more experience and confidence, you can consider expanding your operation.

3. Choose the Right Location: The location of your beekeeping operation can have a significant impact on the health and productivity of your bees. Be sure to choose a location that is sheltered from wind and rain, with access to a source of water and nectar-rich plants. Avoid areas that are heavily trafficked by humans or animals, as this can disturb your bees and increase the risk of theft or vandalism.

4. Provide Adequate Food and Water: Bees need a reliable source of food and water to survive. Ensure that your bees have access to a variety of nectar and pollen sources, and provide a clean source of water near the hive.

5. Monitor the Health of Your Bees: Regular monitoring of the health of your bees is essential to ensure their survival. Keep an eye out for signs of disease, pests, or other issues, and take appropriate measures to address them promptly.

6. Practice Good Hive Management: Proper hive management is essential to the health and productivity of your bees. Regularly check the hive for signs of problems, replace old or damaged equipment, and ensure that the hive is properly ventilated. Inspect your hives at least once a week during the active season, checking for signs of disease or pest

infestations. You should also monitor the amount of honey in the hive and ensure that your bees have enough food to survive the winter.

7. Use Protective Clothing and Tools: Protect yourself from bee stings by using protective clothing and tools, such as gloves, a veil, and a smoker. These tools can help you manage your bees without risking injury.

8. Stay Connected to Other Beekeepers: Join a local beekeeping associations and clubs that provide opportunities for beekeepers to connect with like-minded individuals, share knowledge and resources, and participate in community events. Joining a beekeeping association can be a great way to learn from experienced beekeepers and stay up-to-date on the latest trends and practices in beekeeping.

9. Be Patient: Beekeeping requires patience and dedication. Don't be discouraged by setbacks or challenges, and remember that the health and productivity of your bees will improve with experience.

By following these tips, you can create a successful and thriving beekeeping operation. Remember that beekeeping is a rewarding and fulfilling hobby or business that requires patience, dedication, and a love for these important creatures.

Thank you for buying and reading/listening to our book.
If you found this book useful/helpful please take a few minutes
and leave a review on Amazon.com or Audible.com
(if you bought the audio version).

Table of Contents

INTRODUCTION

Bible Restorative Justice is an approach to justice and conflict resolution that draws inspiration from biblical principles, particularly from the Christian faith. It combines elements of restorative justice with the moral and ethical teachings found in the Bible. Restorative justice, in general, focuses on repairing harm and restoring relationships rather than solely punishing offenders.

In the context of Bible Restorative Justice, several key principles and concepts are often emphasized:

1. Forgiveness: The Bible emphasizes the importance of forgiveness as a central component of Christian faith. Bible Restorative Justice promotes forgiveness as a means of healing and reconciliation. It encourages victims to forgive offenders and offenders to seek forgiveness.

2. Reconciliation: Restoring broken relationships is a fundamental goal. Bible Restorative Justice seeks to facilitate reconciliation between victims and offenders, as well as within the broader community.

3. Repentance and Restoration: Offenders are encouraged to acknowledge their wrongdoing, show genuine remorse, and take steps to make amends for their actions. This can include restitution, community service, or other forms of reparative actions.

4. Community Involvement: The community plays a significant role in the process. It often participates in the resolution and provides support to both victims and offenders as they work towards healing and reconciliation.

5. Healing and Restoration: Bible Restorative Justice prioritizes the healing of the victim and the transformation of the offender. It aims to address the root causes of the wrongdoing and seeks to bring about positive change in the lives of those involved.

6. Restitution: Making amends for the harm caused is an essential aspect. Restitution may involve compensation for damages, but it can also include non-monetary actions that benefit the victim or the community.

7. Accountability: While forgiveness and reconciliation are central, there is still an emphasis on holding offenders accountable for their actions. This accountability is not solely punitive but focuses on personal growth and transformation.